Abstracting Scientific and Technical Literature

ABSTRACTING
SCIENTIFIC
AND
TECHNICAL
LITERATURE

AN INTRODUCTORY GUIDE AND TEXT FOR SCIENTISTS, ABSTRACTORS, AND MANAGEMENT

ROBERT E. MAIZELL
Olin Corporation Research Center
New Haven, Connecticut

JULIAN F. SMITH
Hickory, North Carolina

T. E. R. SINGER
New York, New York

ROBERT E. KRIEGER PUBLISHING COMPANY
HUNTINGTON, NEW YORK
1979

Original Edition 1971
Reprint 1979

Printed and Published by
ROBERT E. KRIEGER PUBLISHING CO., INC.
645 NEW YORK AVENUE
HUNTINGTON, NEW YORK 11743

Printed in the United States of America

Library of Congress Cataloging in Publication Data

Maizell, Robert Edward, 1924-
    Abstracting scientific and technical literature.

    Reprint of the edition published by Wiley Interscience, New York.
    Bibliography: p.
    Includes index.
    1. Technology—Abstracting and indexing.   2. Science—Abstracting and indexing.   I. Smith, Julian Francis, 1893-   joint author.   II. Singer, Tibor Eric Robert, 1902-1966, joint author.   III. Title.
[T10.8.M35 1979]   029'.9'5   78-9756
ISBN 0-88275-703-2

For Elizabeth, Mona and Susan

Preface

This book tells the beginning abstractor and information scientist how to abstract the literature of science and technology. It will also be of value to scientists and to management personnel in universities, industry, business, and government who write and use abstracts or are responsible for the management of abstracting.

We have designed the book so that it can be used effectively either as a text or for self-study. The first pages describe the audiences that constitute the background and the framework within which today's abstractor works and which have a vital impact on his activities. We then explain the management of abstracting operations, specific details of abstracting techniques, and the function of the computer and other modern tools.

Since most of the large, regularly published abstracting services have their own in-house training programs, this book is intended primarily for smaller abstracting operations and individual scientists. However, the material presented here should be of value also to those who are interested in furthering their careers with one of the larger abstracting services.

Although many of our examples are drawn from the well-documented and heavily abstracted field of chemistry, we have written the book so that it is applicable to almost all fields of science and technology. We have covered both the broad principles of abstracting and specific techniques designed to meet the needs of special audiences and special fields.

ROBERT E. MAIZELL
JULIAN F. SMITH

New Haven, Connecticut
Hickory, North Carolina
June 1970

Acknowledgments

For reviewing all or parts of our manuscript, we thank especially the following persons: Professor S. Siggia, Department of Chemistry, University of Massachusetts; Dr. H. E. Kennedy, Biosciences Information Service of *Biological Abstracts*; R. E. O'Dette, Chemical Abstracts Service; Miss Stella Keenan, National Federation of Science Abstracting and Indexing Services; S. E. Furth, International Business Machines Corporation; J. B. Haglind and the late Dr. H. G. Lindwall, both of Olin Corporation; and Dr. B. S. Schlessinger, Southern Connecticut State College.

For permission to reprint certain materials, we thank a number of publishers, as noted specifically throughout the book. Special help was supplied by individuals at *Biological Abstracts*, American Chemical Society Headquarters, Chemical Abstracts Service of the American Chemical Society, Engineering Index, Inc., and the American Petroleum Institute.

Most of the manuscript preparation was ably done by Miriam Chernoff; thanks are also due to Kathleen M. Smith and Mona Maizell for aid and encouragement in the preparation of the manuscript. Susan Maizell provided valuable ideas and added inspiration.

Finally, to our wives, our thanks for their patient understanding, and to Dr. W. E. Hanford, Vice President, Research and Development, Olin Corporation, our thanks for his enthusiastic approval to initiate work on the book.

Our distinguished colleague, Dr. T. E. R. Singer, whose name appears as co-author on the title page, died before the manuscript was completed.

R. E. M.
J. F. S.

Contents

List of Illustrations

Abstracting Scientific and Technical Literature

chapter 1
Introduction

1.1. DEFINITIONS

An *abstract,* simply defined, is a condensation that presents succinctly the objectives, scope, and findings of a document. This information is usually conveyed together with an indexing system, which further helps to identify document content. An abstract, as a rule, is aimed at a specific group of users who either may not read or may not have easy access to the original document. There are different kinds and functions of abstracts (described in Chapter 6 and elsewhere), but the relatively simple definition given here should suffice for now.

From this simple concept stem a powerful scientific tool of great value, many sophisticated (and unsophisticated) information-processing organizations, and the activities of numerous people who produce and use abstracts.

An *abstractor* is an individual who skillfully writes or edits condensations of often lengthy and complex documents. The nature of his work is the unifying theme of this book.

1.2. ABSTRACTORS AND ABSTRACTING ARE IMPORTANT

To scientists, engineers, executives, and information specialists in today's world of technology and science, the work of abstractors has special meaning and importance. Abstractors and their abstracts play a vital role both in coping with the influx of new data and in searching for older data.

1

Abstractors and their abstracts promote current awareness, bridge language gaps, prevent duplication of previous work, and save time and dollars at almost all levels in the majority of organizations. The scientific and technological community has a vital stake in the quality and efficiency of the work of abstractors—work that provides at least a partial solution to the almost universal lament of the scientist: "I just don't have enough time to keep up with the literature."

Abstractors and their abstracts help to conserve research time, hence, research costs, which can be as high as $20 per technical man-hour, or $40,000 per technical man-year (1970 estimate; includes salary, fringe benefits, and overhead costs). Moreover, a research program often involves several people. In some cases, an abstractor can eliminate the cost of an entire research program by indicating that the work has already been done.

Many research and development executives consider abstracting services to be not only essential tools but also real bargains. The average per page subscription of one major published abstracting service costs less than a postage stamp. Many executives would place the real value of such a service at about the salary of a good industrial chemist with a Ph.D. degree (about $20,000 in 1969).

A pilot study (1) of the reading habits of 1394 industrial chemists by the Subcommittee on the Economics of Chemical Information Systems of the American Chemical Society Committee on Corporation Associates illustrates this point. (This is one of the few good studies on the cost effectiveness of information systems.) Based on replies to a questionnaire, it was estimated that the use of abstracting services (averaged over the number of all chemists studied) saves each chemist 5.4 hours per week. Translated into dollars (at $20 per man-hour) this amounts to a savings of about $100 per chemist per week, or $5200 per year. When the latter figure is multiplied by the number of chemists in an organization, the total cost savings is an impressive testimonial to the value of abstracting.

1.3. ABSTRACTING IS BIG BUSINESS

Not only is abstracting a powerful scientific tool of great value, but it is also big business. It involves the efforts of thousands of abstractors, the expenditure of millions of dollars, and the publication of thousands of pages. The range of activities is broad, extending from simple, one-man operations to complex, large organizations.

The growth of abstracting has roughly paralleled the growth of science itself. For example, the total budget of *Chemical Abstracts (CA)* in 1907 was only $15,500 (2). In 1969 the annual budget of Chemical Abstracts Service was $16.5 million, including government supported contract research. On August 5, 1968, *CA* printed its four-millionth abstract, and the five-millionth is expected to be published in 1972. About 3100 men and women around the world abstract part time for *CA,* and there is a full-time staff of more than 1000 at the Columbus, Ohio, offices of Chemical Abstracts Service (CAS). More than 12,000 scientific and technical journals published in more than 100 nations are covered, as are patents issued by about 25 countries. Material for abstracts is translated into English from more than 50 different languages. All of this effort, it has been estimated, will culminate with the printing of 27,114 pages of abstracts in *CA* and more than 20,000 pages of issue and volume indexes for the 249,777 abstracts that appeared in *CA* in 1969.

The Biosciences Information Service (BIOSIS) of *Biological Abstracts* has a regular staff of about 170 in its Philadelphia offices and is an annual operation in excess of $3 million. Approximately 350 volunteer specialists (section editors and abstractors) around the world assist. In 1969, BIOSIS published about 135,000 abstracts in *Biological Abstracts,* and a related publication, *BioResearch Index*, carried an additional 85,000 citations—a total of 220,000. More than 2 million abstracts have been published since this service began in 1926. The two-millionth appeared in the December 1, 1969, issue of *Biological Abstracts,* and the three-millionth is projected for 1973. This recent rapid growth contrasts sharply with the 34 years needed to reach the 1 million mark, and the 8 years and 10 months needed to reach the 2 million mark (3).

These data are all the more impressive when we note that there are hundreds of published abstracting services, and an uncounted number of abstracting services prepared by smaller organizations for in-house use (within the organization). In addition, individual scientists prepare many abstracts for their reports and other activities. For a world guide to published abstracting services in science, see

> Federation Internationale de Documentation, FID Publication 455, Volume I, *Science, Technology, Medicine and Agriculture,* The Hague, 1969. (Also available from the National Federation of Science Abstracting and Indexing Services, Philadelphia.)

This publication includes approximately 1300 abstracting journals and card services and also selected primary journals with important abstracts sections.

Abstracts can originate in a variety of sources, such as governments, trade organizations, professional societies, universities, business firms, independent journals, and individuals.

1.4. COMMUNICATIONS PATTERNS AMONG SCIENTISTS

It is important for the abstractor to be aware of the broad framework within which he works. This framework is the general pattern of communications between scientists. The modern abstractor knows that many scientists obtain much of their information from their colleagues by such informal means as face-to-face conversations, telephone calls, and correspondence.

In addition, scientists acquire much information in more formal ways, such as national, regional, and local technical meetings sponsored by professional societies. Primary journals, review journals, and abstracting services also provide information.

The well-informed abstractor is also aware of the relatively new computerized current alerting and awareness services described later in this book. Some of these services utilize "conventional" full abstracts, but others present merely a title (or a brief note on the content of the document) and the reference.

The sequence of communications between scientists often follows this approximate sequence of overlapping principal steps:

1.   Writing and distributing of reports for in-house use.

2.   Submitting of patent applications (usually requires preparation of an abstract).

3.   Informal exchanging of results and ideas with colleagues in other organizations (face-to-face meetings, telephone calls, correspondence).

4.   Presenting of results at professional society and other technical meetings (usually requires that the author prepare an abstract for use by meeting attendants).

5.   Writing and submitting of papers to journal editors for possible publication (usually requires that the author submit an abstract).

6.   Publication of accepted papers in journals or of accepted patents.

7.   Announcement of publication by current alerting services (express announcement or computerized services are discussed in more detail later in this book).

8.   Preparation of abstracts and of index entries for published papers and issued patents by personnel from one or more abstracting services.

9.   Printing of abstracts and indexes in full size, microfilm, or microfiche form.

10.   Use of abstracts by other scientists.

All of this is part of the evolving communications pattern for the scientist—a pattern that is of great concern to most scientists today. At the beginning of the twentieth century most scientists could read almost all of the journals and patent literature in their fields. When the literature started to increase in volume about six decades ago, scientists established abstracting services to keep up to date. Today, most scientists have time to read only very few journals. And the major abstracting services now cover so much literature that reading even a single issue of one of these services cover-to-cover is out of the question for most individuals. In fact, the scientist who has the time to read all of the pertinent sections of a major abstracting service in his field on a regular basis is rapidly becoming somewhat more the exception than the rule.

This is not to say that abstracts are a disappearing breed. On the contrary, abstracts are very much needed and are widely used for reference and for retrospective searching. Also, many scientists continue to do their own abstracting for their personal files. Moreover, when published together as "packages" in carefully limited fields, abstracts are very valuable in alerting the user to currently available information.

The development of new approaches and new tools, such as the computerized express services, enable abstractors to meet the needs of today's scientists more efficiently than before.

1.5.   ABSTRACTING: ART AND CAREER

The art of abstracting remains as important as ever. Even scientists involved with the newer approaches to information try to take full advantage of the advances made in this art—an art based on more than 60 years of experience and still improving. Similarly, the alert abstractor keeps himself as flexible as possible to meet needs that are both changing and challenging, even as science itself.

Abstracting, like any other art, reaches its ultimate perfection through practice. The techniques and rules used vary, but the basic ground rules of modern theory and practice can be learned by qualified, diligent, and intelligent scientists.

Good abstractors can make significant contributions to the advancement of science. The work is challenging and interesting and offers exceptional opportunities for keeping abreast with advances in science and technology. This is discussed in more detail in Chapter 3.

1.6. READING AND USING ABSTRACTS: LIMITATIONS

Abstractors and other scientists should be fully aware of the possible problems involved in the use of abstracts. To what extent, for example, can the content of an abstract be relied on without recourse to the full original document? This depends in part on the reputation of the abstract journal for consistency (usually well known) and of the abstractor (usually not well known except in smaller organizations). Also to be considered are abstract length and abstract orientation (writing to meet user group needs).

Even the best of abstractors can make omissions because of either personal point of view, or the policy of the abstracting service, or both. An abstractor cannot be expected to cover everything thoroughly—to do so would defeat the purpose of the abstract. Also, most abstractors write in a way that reflects their training, experience, and interests, no matter how objective they try to be. Users should know about this "built-in" limitation of abstracts.

Whether to consult the original document depends primarily on the needs of the abstract user. For example, if the user is a scientist engaged in a highly critical project or in one involving significant expenditures of time and dollars, he should consult many of the originals, using abstracts to narrow his selection. For patent and other legal matters, it is often mandatory to read the full original text. Also, in matters involving human health and safety, there is no substitute for the original, although abstracts can help to focus attention.

On the other hand, a top executive, responsible for large numbers of people and many projects, may of necessity be forced to rely to a great extent upon abstracts, if he is to use his time most effectively.

1.7. USE OF ABSTRACTS AND EFFECTIVE RESEARCH

Many scientists use abstracts as an integral part of their research efforts. Use of abstracts is closely related to the use of other kinds of literature. It takes extra effort and initiative to follow the primary literature in the first place by reading a few key journals. And it takes at least as much initiative and some training to locate and read abstracts. But the dividends can be well worth it.

We think that scientists who have this initiative and are willing to make the effort to use both original literature and abstracts are, as a rule, more effective scientists. According to one study (4), the most creative chemists make more use of almost all forms of technical literature, including abstracts, than the less creative chemists.

If scientists know that management explicitly encourages use of abstracts and other forms of literature, they are more likely to systematically and conscientiously spend time reading in their library or information center.

To make effective use of abstracts (and of other kinds of literature), the scientist must be willing to make an extra effort; he must be aware of both the benefits and the pitfalls inherent in the use of abstracts and must be trained in their proper use; in addition, he needs management support and encouragement. The dividends can be substantial: saving time; avoiding duplication of previous research; building for new work on a sound foundation of facts and ideas; and helping to unlock the gates to creativity.

chapter 2
User Groups, Clients, and Audiences

Whether he knows it or not, the abstractor usually writes for several different audiences or communities of users. It is important for the abstractor to realize this and to have clearly in mind the kinds of user groups that will make use of his efforts. To the extent that he can do this, he can work more productively and effectively. As we explain later (Section 7.7), *abstractors orient their output to meet the special needs of their clients whenever possible.*

In the review of different audiences presented below, we can also see that the work of the abstractor is becoming increasingly important.

2.1. SCIENTISTS

Scientists make up the largest group of users of abstracts of scientific documents. Abstracts are used by all kinds of scientists: young and old; academic and industrial; pure and applied; laboratory and nonlaboratory; managerial and nonmanagerial; and generalists and specialists.

2.1.1. Providing Access to the Literature

Scientists use abstracts in many different ways, and the abstractor needs to keep these uses in mind as he works. The most important use of abstracts is that of providing to scientists access to a vast amount of literature both in specific disciplines (or fields of special interest) and in fields which are multidisciplinary. Most of the other uses noted below overlap with this fundamental use.

2.1.2. Crossing Language Barriers

Abstractors enable the scientist to cross language barriers. Important new data are published in many languages. No scientist can know all languages, but he can gain access to the contents of publications in many languages by using abstracts published in his native tongue, as prepared by multilingual abstractors.

2.1.3. Current Awareness

Abstractors can help scientists to keep up to date. It is obviously much quicker and easier for a scientist to scan an abstracting service in his field than to scan all of the many journals and other publications of interest to him. Thus, the abstractor can perform what is called a current awareness function, which enables the scientist to find valuable items he might otherwise have missed.

But, as noted earlier, the abstractor should be aware that some of the major abstracting services published on a regular basis have become so large that reading all of even a single issue of the service can be a real chore—and is often impossible. For this reason, many *conventional* abstracting services are used increasingly *only* for retrospective searching (except when there is no special current awareness publication in the field).

Associated with the problem of volume is the problem of time lag. Many abstracts appear about three months or more after the original documents are received by the abstracting service—at least this is true with many large services published by professional societies or commercial firms.

To meet the twin challenges posed by volume and time lag, special current awareness services have been developed. These services are discussed in more detail in Chapters 13 and 14.

2.1.4. Retrospective Use

Scientists use abstracting services extensively for searching the non-current literature. The use of conventionally prepared abstracts for retrospective searching is quite heavy—probably much heavier than that for current awareness. Scientists in some fields read (and find useful) abstracts written as long as 20 or even 60 years ago, although ordinarily the last 10 years is a more typical outer limit. Today's abstractors should be aware of this if only to observe how much the quality of abstracting has improved over the years. An even more important consequence of this for today's abstractors is that an abstract is not a transient or ephemeral piece of work. Like a work of art, its value can sometimes increase with the years. This calls for nothing less than the best effort by the abstractor.

2.2. PATENT ATTORNEYS

Patent attorneys are interested primarily in the legal aspect of abstracts. All patent attorneys are familiar with the weekly *Official Gazette of the United States Patent Office;* this is their principal source of information about issued United States patents in abstract form. Almost every other industrialized country also has a gazette covering the patents of that country.

Coincidentally, however, the abstractor should be aware that *scientists* also find patents of great value since they represent a major information source about new technologies. Patents also provide information very rapidly. In particular, the Netherlands, Belgium, and South Africa issue many patents only six months or so after application is made. This is often the first time that details of a new invention are made publicly available. An increasing number of countries have adopted this so-called "quick issue" system. The word "issue" as used in this context means that the patents are open to public inspection; it does not necessarily mean that the patents have been legally granted.

2.3. EDUCATORS

Another major user group is the teacher of science, especially in colleges and universities. The abstract enables professors to keep up to date in their teaching and research without the need for reading many original documents. Both professors and their students benefit.

2.4. INFORMATION SCIENTISTS

Information scientists (consult Glossary, page 279, for term definition) are heavy users of abstracts. Much of their use of abstracts is a service for others and results in such products as (*a*) current awareness notices and bulletins and (*b*) search reports based on study of the older literature (retrospective searching). These services can be performed for the organization as a whole, for specific scientists or departments, or for individual managers, laboratory scientists, and patent attorneys.

The services can be offered on a continuing basis or on a one-time basis; also, they can be requested or unsolicited. These kinds of services are extremely popular and heavily used, especially in industry. Their success depends heavily on abstracts. See Chapter 11 for more details on literature searching.

2.5. SCIENTIFIC-MEETING AUDIENCES

An important medium for rapid communication of scientific results is the meeting or symposium. The audiences for abstracts of these papers include management, those who attend the meetings and symposia, and the symposium or meeting-program chairman. The writer of the paper writes his own abstract.

Management is usually the first to review such abstracts, as submitted by scientists in their organization, to ensure that the work to be presented is of high quality and also that it does not reveal data prematurely. Attendants at meetings or symposia are often influenced by premeeting abstracts in their selection of papers that they plan to attend or for which they will request reprints at a later date. The meeting or symposium chairman may accept, reject, or modify a proposed paper based on the abstract. Thus abstracts play a vital role in this highly significant medium for communicating new information.

The procedures of many scientific and technical societies for their meetings require prospective authors to submit abstracts three to four months before the meeting. For this reason, abstracts are usually written before the full paper to obtain advance approval and acceptance from (*a*) management and (*b*) meeting chairmen. After all, there is no point in writing a full paper until the abstract is approved.

From management's viewpoint, the author should give enough information so that a valid judgment can be made. The decisions that management needs to make are whether the work is substantive, whether it reflects credit on the organization, and whether the information is satisfactory from the legal (especially patent) and proprietary standpoints. Legal, public relations, administrative, and research and development personnel are among those who may take part in the screening process.

The interest of the meeting or program chairman in meeting-paper abstracts is similar to that of management in that he wants to ensure a high-quality program of current interest.

2.6. ABSTRACTS FOR JOURNAL PAPERS; AUTHOR ABSTRACTS

Abstracts written by scientists to accompany journal papers usually have management as the first audience. Approval is needed before the full paper is written, for the same reasons that abstracts of scientific-meeting papers need approval. These abstracts are then looked at by journal editors and referees who pass on the acceptability of both abstracts and papers, and, ultimately, by the readers of the journal. Emphasis on the reviewing of abstracts as a part of the refereeing-editorial process is relatively recent and is a significant step in the right direction.

Another extremely important audience of journal paper abstracts is made up of abstracting-service personnel. The current trend by these personnel is to rely more on these "author abstracts"—so called because they are written by the same person who writes the original article. Author abstracts can offer abstracting-services personnel important advantages in reducing the time lag between journal publication and abstract and index publication, and also can offer obvious advantages in economy even though editing (see Section 3.3) may still be needed. An optimum situation is one in which the author abstract can be taken with no change directly from the galley or page proof of the journal. This is an especially notable case of the abstract assuming increasing importance. There are, however, certain problems in the use of author abstracts; these are discussed in Sections 3.7.2 and 6.7.

2.7. MANAGEMENT

We have already noted that management reads abstracts to approve (or disapprove) presentation and publication of papers. Management (executives) also reads abstracts that appear at the beginning of in-house reports—and reports are part of the lifeblood of the decision-making process in most organizations.

This use is especially important because executives frequently do not have the time to read complete reports. An abstractor who writes well can save the executive much needed time and make decision-making easier and more accurate. Since executives control purse strings and make vital decisions, the writing of abstracts for their use, especially in in-house reporting, is worth extra time and effort.

2.8. LIBRARIANS

Many librarians subscribe to abstracting services rather than to the thousands of individual journals covered by these services. To subscribe to all of the original journals, even in one or two fields of science, could lead to space problems and, ultimately, to the need for construction of costly new library buildings. The same reasoning applies to other original documents covered by abstracting services, such as technical reports, patents, and trade literature, all of which can take up many linear feet of shelving. Thus, if the librarian can subscribe only to those journals of greatest interest (assuming that most of the rest will be covered by abstracting services), many dollars can be saved. But these savings do not apply equally to the larger libraries, which subscribe to almost everything in their fields of interest.

2.9. INDEXERS

Indexers are major users of abstracts. (The indexing of abstracts is discussed in detail in Chapter 9.) Their work is very important, since the index is essential in guiding the scientists to the abstracts. The indexer and abstractor must work together very closely; each contributes something to the identification of document content. Not infrequently, indexing and abstracting of documents are done by the same person.

chapter 3
Managing Abstracting Operations

3.1. HOUSING THE ABSTRACTOR

Since the work of the abstractor is intellectual in content and nature, he must work in a comfortable, quiet, and private office of adequate size (100–120 sq. ft.). Sharing an office with another abstractor is usually not satisfactory, especially if both abstractors use the phone or have visitors. But the abstractor should not be secluded and isolated from the rest of the organization. For working at home, the abstractor should have a private study, free from distractions. Lighting must be adequate—100 foot-candles is a reasonable level. Also, a smooth, nonreflective reading and writing surface is highly desirable.

Certain basic reference tools should be within arm's reach. These include one or two handbooks such as the *Handbook of Chemistry and Physics,* English language and foreign (bilingual) dictionaries, and one or two advanced textbooks in the fields in which the abstractor most frequently works. Ready access to a research library or information center is essential for many reasons; not the least of these is to help the abstractor to keep up to date. In addition, abstracting guidelines (see Section 3.10), including lists of acceptable abbreviations, should be on the desk top.

3.2. EQUIPMENT AND WORK HABITS

The work habits of the abstractor will indicate whether any additional equipment is needed beyond the bare minimum requirements of a chair and a desk or table. For example, some abstractors will want a typewriter for preparation of their copy, especially if they do not prepare a preliminary draft.

If adequate secretarial assistance is available, the abstractor may want to write his abstract in longhand for subsequent typing by a secretary. This procedure requires no special equipment except, of course, a typewriter equipped with the more common scientific symbols. Use of devices such as the "Typit" attachment, with the standard typewriter keyboard, adds the flexibility necessary in working with scientific material. "Typit" (made by Mechanical Enterprises, Alexandria, Virginia) is a set of removable keys that can be quickly inserted as needed. Or an IBM "Selectric" Typewriter (with interchangeable typing spheres) can be used. Either of these, alone or in combination, can provide the capability of using a large variety of scientific and technical symbols.

Other abstractors prefer to dictate a first draft into a recording unit for later typing by a secretary. Since most people can talk several times faster than they can write, the savings in time to both abstractor and secretary can be significant. Another advantage that some abstractors find in dictating is in overcoming inertia in getting their thoughts on paper.

Dictating is a skill that requires practice and preparation by both the abstractor and the secretary. The abstractor must have his material and thoughts well organized, enunciate clearly, provide concise transcription directions, and spell out (letter by letter) anything unfamiliar such as long proper names and unusual technical words. The use of a standard working form can be helpful in ensuring that all of the elements of the abstract entry are complete and properly recorded.

The secretary, if she is to function with maximum efficiency, must learn the mechanics of using the transcribing machine and the basic nomenclature of the field, including the spelling of frequently occurring technical words.

One of the best discussions of dictating and its advantages and disadvantages will be found in *Chemical Engineering* (5), (6).

If first drafts are typed by secretaries, and if the abstracting operation is a relatively large one, lease or purchase of units such as the IBM "Mag Card Selectric® Typewriter" or the "Friden 2340" should be considered. These units can prevent retyping of the entire abstract by "capturing" the draft on either a magnetic card or punched paper tape. The original typing can then be "saved" and automatically retyped at high speed (about 20 characters per second) after any necessary corrections are made.

Some abstractors now have available to them more sophisticated means of input made possible by developments in computer technology. This is discussed in Chapter 14.

3.3. PROOFREADING AND EDITING

Abstracts need careful proofreading and, very often, editing. Proofreading is especially important if the abstractor prepares a first draft from which a final copy is typed by a secretary and then later typeset—the very nature of this process allows multiple chances for error.

It is especially important that the complete reference to the original source document be proofread carefully. The title of the journal, volume and page numbers, and the exact spelling of authors' names should be checked in accordance with guidelines (see Section 3.10). This kind of proofreading makes it possible for users to locate an original source accurately. Also, numerical data in the text of the abstract should be carefully proofed since typographical errors are most likely to appear in this kind of material.

If time and money permit, the abstractor may want to edit (revise) the draft of his abstract as necessary. In larger organizations there may be a special group or an individual responsible for the editing of abstracts and also additional proofing steps. Editors concern themselves with such matters as general conformity to style (e.g., abbreviations), grammar, spelling, and scientific correctness. They may also do some indexing.

Author abstracts, in particular, may need editing. An author abstract written by a scientist whose native tongue is not English deserves special editorial attention in any abstracting operation intended primarily for an English-speaking audience.

It is clear that chances for error are *increased* by multiple transcription and multiple keyboarding (typing). Chances for error are *decreased* if repetitive transcription and keyboarding can be kept to a minimum; time and money are also saved.

3.4. SELECTION OF ABSTRACTORS

3.4.1. Criteria

What kinds of scientists make good abstractors? The following criteria might be used:

1. Intelligence.
2. Up-to-date knowledge of relevant fields in which the abstractor will work and of the significance of new developments in these fields.
3. Ability to read and write quickly, accurately, and clearly. A flair for the correct choice and use of words.
4. Ability to read at least one major foreign language, and ability to learn others as needed.

5. Sufficient imagination to find valid shortcuts.

6. Ability to work independently, but to know when to call for help from others when needed.

7. A general orientation toward the literature, and a special interest in the literature.

8. Neat and well-organized work habits.

9. Reasonable ability to get along with all kinds of people at all levels —an essential for almost any job.

Scientists who have most of these capabilities can usually learn abstracting, even with no previous experience with abstracting.

Of the several skills and traits mentioned above, one of the most important for the abstractor to master is the ability to write well—the ability to marshal facts and to summarize them clearly and concisely in a coherent form. A good abstractor already has (or should strive to develop) a concise writing style, which saves time and money for all concerned— abstractor, editor, typist, printer, and abstract user. But the abstractor should not have a style that is so concise or stilted as to obscure clarity. Nor should his style be so rigid and inflexible as to be unable to handle the great variety of original documents with which he is likely to come into contact.

3.4.2. Selection Process

Abstract writing skills can be judged in part by studying samples (if available) of the prospective abstractor's writing and evaluating these against the criteria suggested above. If sample abstracts are not available, correspondence and reports can give clues as to writing ability.

If it is desired, formal tests can be used as a further aid in selection. These tests could cover one or more of the following:

1. Knowledge of specific fields of science.
2. Knowledge of terms or nomenclature in specific fields of science.
3. Knowledge of foreign languages.

3.5. ABSTRACTOR SPECIALIZATION—THE GENERALIST ABSTRACTOR

Another important consideration in the selection of abstractors is field of specialization. Even in subfields of science, abstractor specialization is an ideal situation. For example, a chemist who is a specialist in polymer science can probably do a good job of abstracting the literature of polymer chemistry. He may have a little difficulty if he tries his hand at abstracting literature in other fields of chemistry, and he will almost certainly have some trouble in doing a fully adequate job in a field such as electrical engineering. In small or one-man operations, however, there is often no choice. A fully qualified scientist, working in a small organization (or by himself), may need to work in relatively unfamiliar fields simply because there may be no one else around to do the job.

Since information scientists (scientists who work full time in an information center), for example, are almost always called upon to do some abstracting in unfamiliar fields, they become generalists—at least in this activity.

Let us amplify the points just made. If the abstractor is keenly intelligent and perceptive, and can write well, he can prepare reasonably satisfactory abstracts of scientific material in some fields other than his own. But if he is to function effectively as a generalist, he will need to spend time in doing continuing background reading in the various fields in which he abstracts. Also, if he is a beginner, qualified specialists should review his work in these various fields during a training period and for some time thereafter.

3.6. FOREIGN LANGUAGES AND THE ABSTRACTOR

An abstractor who has a reading knowledge of one or more of the major foreign (non-English) languages, and who can learn additional languages with relative ease, has a most valuable asset. This is especially true if the abstractor will work in fields in which much important work is published in languages other than English.

Recent trends have lessened the importance of the linguistic ability of the abstractor. Many foreign journals publish author abstracts in more than one language; one of these languages is usually English. Other foreign journals contain full articles in English or provide a virtually complete translation. An example of the latter case is the German journal on plastics, *Kunststoffe,* in which each issue contains both the original German and also English translations of important articles. Also, cover-to-cover translations are available for many Russian journals, especially in the fields of physics and chemistry; the American Institute of Physics has been a leader in this effort.

Even so, the abstractor should try to keep his linguistic capabilities sharp. And he should be able to use bilingual dictionaries with facility. Skillful use of graphs, tables, equations, and the like also aids the abstractor in reading foreign language material; graphics are a kind of universal language.

3.7. SOURCES OF ABSTRACTORS

3.7.1. For Large Published Services

Part-Time Abstractors. Some of these services rely heavily on part-time abstractors who are often scattered throughout the world. This is made necessary by such factors as volume of work, diversity of field, and (primarily) diversity of language of the original documents. A prerequisite is the identification and recruiting of scientists willing to do this kind of work. The professional advantages to the scientist are noted on page 20.

The hoped-for benefit to the abstracting service is a combination of both field and language expertise. Also, costs can be low since the fees (if any) paid for such part-time services are often minimal. Sometimes, the work is done as a professional courtesy.

Possible disadvantages include difficulty in getting part-time people to complete and return abstracts promptly. Also, there may be editorial problems involving both format and correct use of the English language, especially with abstractors whose native tongue is not English.

On-Site. A staff of full-time, on-site abstractors can be employed, either as a partial substitute for part-time, off-site people, or as a complete replacement. Full-time abstractors have a potential advantage in being able to follow an abstract through all of the various stages of its production, including selection of documents worthy of abstracting, abstract preparation, proofreading, and indexing. Full-time abstractors are especially useful in the indexing process since they become quite familiar with the content of the document in the course of the preparation of the abstract. Also, editorial problems can be minimized if the abstractors are on-site, because of the ease of training and of exchange of information between the editor and his staff.

A principal disadvantage can be the lack of abstractor specialization in some fields—some may need to be generalists. Another disadvantage can be limitations in language capabilities.

3.7.2. For In-House Abstracting

Many industrial organizations do their own abstracting for their in-house bulletins; some of the abstracting can be done by the bulletin editor who often also functions as an information scientist with other related duties in an information center.

For in-house abstracting, management should also consider the use of in-house laboratory scientists as part-time abstractors working through and with the full-time scientific personnel in the information center. For best results, the abstracts flow through an information scientist who acts as coordinator, advisor, troubleshooter, and gentle prodder. This can be especially helpful when the information center staff is not large enough to do the whole job.

There are other advantages. Laboratory people can bring to the abstracting effort a more specialized and critical approach than the information scientist who is usually more of a generalist, especially when it comes to abstracting. Laboratory people are also more able to point out in their abstracts why new material is significant, particularly in their own organization.

One probable difficulty with part-time laboratory personnel as abstractors is that they may need editorial or writing assistance. Another possible disadvantage is that many laboratory people are, logically, primarily concerned with their own research. Abstracts may therefore be late and may need to be coaxed out of the laboratory.

In any effort to involve laboratory people, the complete support of management is needed. Qualified, cooperative, and prompt abstractors are quickly identified from within the ranks of laboratory people by evidence of interest in the literature, writing ability, and expressions of interest.

If management decides that laboratory scientists can be used as abstractors, it may be necessary to convince some of these potential abstractors of the many benefits, which include the following:

1. Abstracting offers a means of continuing one's education by keeping up to date in a formal way. Since continuing education is of great interest to many scientists, this should be a cogent argument.

2. If the work is properly supervised, it should enable the laboratory scientist to become a better writer.

3. Abstracting adds variety to other work.

4. The laboratory scientist should find that his habit of using the literature is reinforced and encouraged—and this is a habit that is useful and important to good research.

Related arguments can often be used to recruit those wives of laboratory men who have technical degrees and who have the time to work at home.

Thus the decision to call on laboratory personnel to abstract depends on: the availability of specialists and generalists to do the various kinds of jobs needed; the willingness and ability of these people to do the job; and the availability of the necessary funds and time. With the increasing importance of the author abstract (especially in journals), there are those who believe that the question may, in time, prove to be academic since the need for auxiliary personnel may no longer exist.

3.8. NONSCIENTISTS AND ABSTRACTING

Common sense dictates that only qualified scientists should be used for the abstracting of scientific literature. The nonscientist may be able to aid in a few limited areas, such as assisting in proofreading, especially with the reference to the original document, and he may also be able to help in the selection of key words for indexing. But the abstract itself should be written by a person who understands the content of the document. Only a qualified scientist (with at least a bachelor's degree) can do this.

3.9. TRAINING

3.9.1. Formal Training

Some new abstractors may already have had training relevant to abstracting in their undergraduate or graduate studies. Most relevant (assuming, of course, that the person already has the necessary scientific training) would be formal courses in abstracting and indexing. Courses on the techniques of technical writing and editing, and on the use of the technical literature, are also helpful.

For those new abstractors who have completed their full-time education, short courses or conferences relating to writing and technical information are helpful. The open meetings and training programs of the National Federation of Science Abstracting and Indexing Services are well worth attending, as are selected short course offerings and conferences sponsored by such organizations as the American Management Association and the American Chemical Society.

3.9.2. On-the-Job Training

Most new abstractors cannot be expected to produce top-notch work at first. Their work, for at least the first few weeks and months, needs regular review and, for the next 6 to 12 months thereafter, less frequent but continuing review. The review process can be achieved with the cooperation of more experienced abstractors. Also, some techniques and skills can be learned if the abstractor is willing to make a special effort. One example is efficient reading (7); another is the art of dictating abstracts into recording machines.

In a small or one-man abstracting operation, the users of the abstracts should be asked to provide the feedback and criticism that are necessary in developing the skills of the beginning abstractor.

Abstractors in any organization, even though they may work in private offices, should be encouraged to spend as much time as possible mixing with their clientele. These stimulating contacts are essential if the abstractor is to meet real needs and keep in complete harmony with the real world.

3.10.   GUIDELINES, INSTRUCTIONS

These are closely related to training. If a new abstracting service is to be established, guidelines should be written down *at the outset*. It is very important to document policy decisions to avoid later confusion.

Guidelines can be set by the abstractor himself, by his users, by his management, or by a combination of these. In larger or more formal organizations, guidelines are usually presented to the abstractor in the form of a booklet with a title such as "Instructions to Abstractors."

For most abstractors guidelines of some sort are helpful. Guidelines are especially useful when there is more than one active abstractor in the same organization. Guidelines then help to ensure and to *maintain* consistency among current personnel despite the turnover that inevitably occurs in most organizations.

Guidelines can be especially useful to part-timers and to beginners; both categories of individuals need much more assistance than full-time and more experienced people.

The guidelines can be very lengthy and detailed. We would expect this to be true especially in large published services that have many abstractors. Because of the large number of abstractors and the great variety of their backgrounds, specific and detailed instructions are necessary. Another reason is that in large organizations many of the abstractors are not located on-site (many are overseas) and therefore must get most of their training through such written instructions. The *Directions for Abstractors* published by Chemical Abstracts Service is especially good, and most abstractors, regardless of subject field, would profit by reading it.

What should a guidelines pamphlet contain? Typically it would cover such topics as the following:

1.   General instructions.
2.   Style.
3.   Special instructions for abstracting various subfields or for abstracting of special kinds of materials (e.g., patents).
4.   Abbreviations and symbols.
5.   Aspects to be emphasized; aspects to be deemphasized or omitted.

Guidelines should be kept flexible and should be modified (in writing) to meet changing needs. A looseleaf notebook arrangement makes this possible. In addition, if it is lengthy, the guidelines manual should have an index and a table of contents.

Among other objectives, guidelines should enable the abstractor to achieve conciseness. This is usually done by encouraging the skillful use of tightly worded sentences that enable the abstractor to present maximum detail in minimum space. Flowery rhetoric has no place in the abstracting of scientific documents. Space can also be saved in other ways, such as by effective use of abbreviations that are well known and accepted in the specific field of science being abstracted (see for a word of caution, Section 7.3).

A further comment on guidelines: if these are too detailed, they may unduly restrain the abstractor and slow down his production of abstracts. A certain amount of flexibility is necessary, particularly to accommodate the differences in writing styles among abstractors. After all, the essential goal of the abstractor is to convey clearly and succinctly the contents of the original document. But abstracts prepared for computer processing may need to follow a very specific, rigid format. (See Chapter 14.)

Figure 3.1 illustrates the guidelines used by *Biological Abstracts*. This figure provides valuable information on the preparation of abstracts and is worth careful study. Also, the American National Standards Institute is expected to issue a standard on abstracts in the near future.

Much of the time the abstractor will find it possible to work within the framework of the guidelines set by others (or by himself). But three basic principles should come before and override all others:

1. The needs of the user are paramount and should be reflected in the abstract.

2. The abstractor must be able to work comfortably within the framework of the guidelines.

3. The abstractor must face realities and do his best to work within any reasonable ground rules imposed by others, but he should also attempt to work for continuing improvements and changes in the guidelines as warranted.

If these conditions are satisfied, the abstractor is more likely to be a productive individual and also is more likely to achieve professional satisfaction.

3.11. COST AND SPEED OF ABSTRACTING

The cost of abstracting is highly variable. Some of the factors involved are:

Fig. 3.1. Example of guidelines for preparation of abstracts. *Source. Biological Abstracts* (BIOSIS). Reprinted with permission.

GUIDE TO THE PREPARATION OF ABSTRACTS
FOR BIOLOGICAL ABSTRACTS–1970

An abstract should be a non-critical, informative digest of the significant content and conclusions of the paper, not a mere description. It should be intelligible in itself, without reference to the paper, but it is not intended to substitute for it. It should be brief (preferably less than 3% of the original, usually 150-300 words in length), written in whole sentences, not telegraphic phrases.

CONTENT

Include:

1. Objective of the study.
2. Name of organism(s) used. Supply both the scientific and common names, if given in the article. For plants, cultivar names may be important.
3. Materials, methods and apparatus and their intended use; when new, describe briefly.
4. New techniques, their uses and characteristics.
5. Specific drugs (generic names preferred) and other biochemical compounds. State the manner of use and route of administration.
6. Principal results and significant conclusions. Always include new or verified data of permanent value such as absorption spectra, chromosome numbers, life cycles, parasite hosts.
7. New theories, new terminology, interpretations or evaluations stated concisely.
8. Definitions of new terms and meaning of special abbreviations used.
9. New taxa, new distribution records (for systematic data see additional instructions below).

For list of "Acceptable Abbreviations" see Fig. 7.2.

RULES FOR ABBREVIATIONS

1. In general, use only those abbreviations listed under "Acceptable Abbreviations."
2. Use the abbreviations for units of weight, measure, time, radioactivity, etc., and the symbols %, o/oo only with figures, e.g., "10 mg" but "several milligrams"; "5% level" but "per-cent of gain."
3. Use figures for numbers except at the beginning of a sentence and when a number is part of a name, e.g., "two-spotted mite."
4. Abbreviate names of chemical elements when used alone or in common inorganic compounds, e.g., "K deficiency" and "NaCl excess" but "potassium 2-naphthyl sulfate excretion."
5. Abbreviate long substantives used repeatedly, such as names of compounds, hormones, enzymes, but only after they have been spelled out the first time they are used in each abstract, followed immediately by the abbreviation in brackets: "succinic dehydrogenase [SDH]," "pollen mother cell [PMC]."
6. Do not abbreviate geographic names (except USA and USSR) and three-letter words.
7. In the citation, abbreviate journal titles as given in the "BIOLOGICAL ABSTRACTS List of Serials with Title Abbreviations." If not available, write out the full names.

Omit:

1. Information contained in the title.
2. Additions, corrections or any information not contained in the original published paper.
3. Tables and graphs and direct references to them.
4. Detailed descriptions of experiments or organisms.

FORM

1. Begin with a citation in the following form: DOE, JOHN A. (Dep. Biol., State Univ., Philadelphia, Pa., USA) and RICHARD ROE. Phosphorus metabolism in the rat. J BIOPHYSIOL 37(4): 152-165. 1970. Use the first author's most recent address, if given. If the title is not in English, give the original title (not underscored) and follow the convention of the language, if known. The translated English title should follow enclosed in brackets [ ]. Capitalize the first letter of the first word of the title and all proper names. Upper case all of the letters in the authors' names and the journal title.

2. When a BIOSIS citation form is available, insert the bibliographic information in the appropriate spaces following the above instructions, but do not underscore the title.

3. Type (preferably) or print the abstract double-spaced on one side of the paper.

4. Avoid colloquial terminology and trade names. For chemicals, including certain abbreviations, use standard rather than proprietary names (see "Acceptable Abbreviations"). For drugs, use the generic name (when given) in preference to the proprietary or chemical name. Registered proprietary names are always written with an initial capital. For example, Levanil is a proprietary name for ectylurea, the generic name for (2-ethylcrotonoyl) urea.

5. Always underscore the Latin names of genera, subgenera, species, subspecies, varieties and forms. Never use the specific epithet of an organism without its accompanying generic name (which may be abbreviated if first given in full), e.g., Escherichia coli ... E. coli.

SYSTEMATICS (Additional Instructions)

CONTENT

1. Include: new or revised taxa of any rank; geographic locations; significant anatomical and physiological characteristics; life-cycle data; hosts of parasites; etc. When many new taxa are given, the number in particular categories may be cited, e.g., "17 new species and varieties in 6 genera of Fungi Imperfecti are described."

2. Distribution data should mention, when given, first records, new localities, maps, range changes, faunistic or floristic changes, and epoch or period of fossils.

3. Include important facts other than the formal, strictly taxonomic information. Summarize or mention new data in any biological field—ecology, evolution, morphology, cytology, etc.

FORM

1. Write the name of an established generic, subgeneric or supergeneric taxon (family, order, etc.) with an initial capital letter, e.g., Cichlidae. If newly proposed, write the names of these taxa all in capital letters, e.g., "The new family BRAUNIDAE is proposed." The name of a species, subspecies, variety or form is never capitalized.

2. Underline the Latin name of an established genus, subgenus, species, subspecies, variety or form with a straight line, e.g., Philornis aitkeni. If newly proposed, use a wavy line, e.g., Philornis downsi sp. nov., ATLANTOPSIS canariensis gen. et sp. nov.

3. For transfers or changes of rank, write the proposed combination as above and, in parentheses, the original name (basonym) and its author, e.g., Xylobium pickianum (Bifrenaria pickiana Schlechter). For changes in status (validity, synonymy) write the valid name, its author if not new here, and, in parentheses, the invalid name and author preceded by an equal sign, e.g., Erisma floribundum Rudge (=E. parvifolium Gleason), or GANELIUS (=Nagelius Benesh preocc.).

1. Salary of the abstractor. This, in turn, depends on such factors as the following:
   (a)   Kind of organization (academic, government, or industrial).
   (b)   Location (some parts of the country have higher wage rates).
   (c)   Education and experience of the abstractor.
   (d)   Abstractor competence, skill, and productivity.

Some occasional part-time abstracting is done without charge as a courtesy on the part of the abstractor.

2. Type of abstract. Critical abstracts are the most expensive, followed in descending order by informative and indicative abstracts. The extent to which author abstracts are used is also a major factor in cost. (See Chapter 6 or Glossary for definitions of abstract types.)

3. Type of document. Factors involved include:
   (a)   Length.
   (b)   Language (English or foreign).
   (c)   Complexity and degree of sophistication.

4. Work habits and practices of the abstractor. Factors involved include:
   (a)   Extent of rewriting (is a draft copy prepared first?).
   (b)   Extent of editing (if any).

The more rewriting and editing, the higher the cost—but also the better the abstract in many cases.

5. Overhead and other costs. Factors involved include:
   (a)   Administration.
   (b)   Typing.
   (c)   Office space.

Depending on the extent to which these factors come into play, we estimate that cost per abstract could be in the range of $15–30, including intellectual effort (writing, editing, proofreading) and typing. This range is cited to indicate that abstracting can be relatively expensive. The actual figure in a specific case could be less or it could be greater, because of variations both in abstracting policies and in accounting practices in different organizations. But it is vitally important to recognize that the cost of abstracting is a figure that can be controlled by decisions and standards set forth by the abstractor or his management. Thus, for example, an organization that demands lengthy abstracts and multiple keyboarding can expect higher costs; an organization which is satisfied with briefer abstracts and can achieve minimal keyboarding can expect lower costs.

How can the abstractor judge for himself whether he is working fast enough? The answer is highly variable and depends on most of the same factors that affect costs.

A hypothetical abstractor should be prepared to write, on the average, about 8 abstracts per day for original documents that do not contain author abstracts. This figure does not include proofing or editing and is, as we have said, subject to wide variation.*

The abstractor can use time as a criterion in evaluating the adeptness and diligence of his performance, especially if he abstracts on a part-time basis. By setting standards for himself, the abstractor can prod himself into getting his work out. The abstractor should strive to complete all abstracts within 48 hours after the receipt of the original document. Such a standard should be set with the realization that it can be affected by total work load and overriding priorities. On the other hand, the demands of his organization may be such that 48 hours may be too slow and that the abstracts may need to be turned out within 8 hours (or even less) after receipt of the original document.

3.12. DUPLICATION OF ABSTRACTING EFFORT

Abstracts of the same document will often be published in more than one abstracting service. One of the reasons for this is exchange agreements. Some of the major published services in different countries have agreements under which each covers some sources less readily accessible to the others. Each duplicates, in its own language, some abstracts from otherwise missing source material. A similar kind of agreement can also obtain between abstracting services in the same language because of differences in accessibility to the original documents. Thus, exchange agreements achieve broader coverage and save funds. (See also Section 14.8.)

Unfortunately, exchange agreements are not always possible. One reason could be a marked difference in *broad* interests (orientation) between the abstracting services involved. Other reasons could be the need to keep interests confidential, to achieve speed of production not otherwise possible, or to orient (see Section 7.7) abstracts to meet *very specific in-house* needs. It is important that both abstractor and abstract user know that these kinds of situations exist.

Nevertheless, cooperative efforts can minimize overlap or duplication. Thus in 1970 Biosciences Information Service of *Biological Abstracts,* Chemical Abstracts Service, and Engineering Index, Inc. announced a detailed review of the relationship between their publications and services.

---

* Some organizations would consider 8 abstracts per day as a bare minimum.

One of the purposes of the joint study is to eliminate any unnecessary duplication within the three organizations.

3.13. PROFESSIONAL ORGANIZATIONS

In the United States, the National Federation of Science Abstracting and Indexing Services (NFSAIS) fosters cooperation among nonprofit scientific discipline-oriented services. The Executive Director at this writing is Miss Stella Keenan and the president is Kenneth C. Spengler, American Meteorological Society. Figure 3.2 provides significant details about the aims, history, special projects, publications, and membership of NFSAIS.

<u>NATIONAL FEDERATION OF SCIENCE ABSTRACTING AND INDEXING SERVICES</u>

<u>PHILOSOPHY OF SERVICE</u>

The National Federation of Science Abstracting and Indexing Services believes that the abstracting and indexing services of the United States are providing the American scientist with good (usually superior) coverage of the world's scientific literature. This has been accomplished by the individual efforts of the independent abstracting and indexing services of the United States, rather than by a large central, monolithic governmental organization. The Federation further believes that, through the cooperative efforts of its members, even greater advances can be made in improving the abstracting and indexing of the international scientific and technological literature, thereby providing better access to the ever-increasing store of scientific knowledge.

<u>BASIC OBJECTIVE</u>

The Federation is endeavoring to coordinate the cooperative work of the various member services, and to seek new ways of improving them. It is also encouraging the development of abstracting and indexing organizations for those specialized subject fields not presently covered by existing services.

The ultimate goal of the Federation is to improve communication among scientists through the documentation (abstracting, indexing, and analyzing) of the international scientific literature.

<u>USER AUDIENCE</u>

Staff of member services, and all personnel engaged in abstracting and indexing work. Many of the Federation's activities and publications are of interest to scientists, technologists, information specialists and librarians.

Fig. 3.2. Basic information on National Federation of Science Abstracting and Indexing Services. *Source.* National Federation of Science Abstracting and Indexing Services, Philadelphia, Pa., Technical Report No. 1, October 1969. Reprinted with permission.

COVERAGE

This is best shown by the statistics on member service publications given in the following table.

PUBLICATIONS AND SERVICES

A Guide to the World's Abstracting and Indexing Services in Science and Technology. 1963
News from Science Abstracting and Indexing Services. Bi-monthly newsletter
Technical Report Series

Proceedings

Conference of American Scientific and Technical Abstracting and Indexing Services.
Philadelphia, January 29-31, 1958. July, 1958.
2nd Annual Conference of NFSAIS
Washington, D.C., February 26-27, 1959. July, 1959.
1962 Annual Meeting of NFSAIS
Boston, Massachusetts, March 28-30, 1962. January, 1963.
1963 Annual Meeting of NFSAIS
Washington, D.C., March 20-22, 1963. November, 1963.
(This volume contains the National Plan for Science Abstracting and Indexing Services prepared for the Federation by Heller Associates in 1963).
1969 Annual Conference Proceedings
Raleigh, N.C. To be Published Fall, 1969.

Seminar Series

Indexing in Perspective - three day course.

Address:

National Federation of Science Abstracting and Indexing Services
2102 Arch Street
Philadelphia, Pennsylvania 19103 USA

Fig. 3.2. *(Continued),* See also page 30.

Information service companies have formed an Information Industry Association to represent their organizations and interests. This group includes about 50 organizations that provide abstracting and other information services. Objectives of the Association include: exchange of information and ideas; the development of standards and policies relating to orderly development of new techniques in the dissemination of information; distribution of information to members and professional organizations; and fostering public relations. The president of the organization, at this

NATIONAL FEDERATION OF SCIENCE ABSTRACTING AND INDEXING SERVICES

MEMBER SERVICE STATISTICS

1957-1969

	1957	1962	1967	1968	Estimate 1969
Abs. of Photo. Sci. Eng. Lit.*	----	2,613	3,593	3,085	3,600
American Geological Institute	----	6,000	11,450	17,029	30,000
American Petroleum Institute*	----	----	29,000	28,800	30,000
+ Applied Mechanics Reviews	4,245	7,200	8,802	9,426	10,000[b]
+ BIOSIS	40,061	100,858[a]	125,026[a]	214,000[b]	220,000[b]
+ Chemical Abstracts Service	101,027	169,465[a]	239,481[a]	251,884[c]	282,000[c]
Computer and Control Abstracts[d]	----	----	6,204	7,311	16,000
Electrical & Electronics Abstracts[d]	6,451	15,038	24,039	30,438	34,000
Engineering Index, Inc.	26,300	45,000	56,560	61,231	65,000
Esso Research & Engineering Co.*	25,000	35,000	10,500	5,300	4,500
Information Science Abstracts*	----	----	1,327	1,570	2,000
+ Mathematical Reviews	9,200	13,382	17,141	15,179	17,000
Medical Documentation Service*	----	----	1,692	1,901	1,800
+ Metals Abstracts[e]	8,219	11,542	23,800	23,007	24,000
+ Meteorological & Geoastrophysical Abs.	5,000	12,000	9,000	9,269	9,000
Oral Research Abstracts*	----	----	6,681	7,256	7,600
Physics Abstracts[f]	10,001	24,236	40,788	50,477	65,000
+ Psychological Abstracts	9,074	8,776	17,202	19,586	20,000
Tobacco Abstracts	1,798	2,510	2,966	2,739	3,000
TOTAL:	246,376	453,620	635,252	759,488	844,500

Notes:

+   Founder member

*   New member - 1968-69

a   Major publication only; does not include subsidiary publications

b   Includes BA and BioI

c   These data include duplicate patents referenced to the abstract in CA through the CA Patent Concordance. In 1958 there were 19,180 such references and in 1969 these will be an estimated 30,000.

d   IEEE joined the Federation in 1969. They became joint publishers with the Institution of Electrical Engineers (U.K.) of EEA in 1968 and of CCA in 1969 (formerly Control Abs).

e   In 1967 Review of Metal Literature (U.S.) merged with Metallurgical Abstracts (U.K.).

f   Figures supplied by the American Institute of Physics, who joined the Federation in 1968.

Fig. 3.2. *(Continued)*

writing, is W. T. Knox, McGraw-Hill, Inc.; Paul G. Zurkowski is the executive director. Member organizations are listed in Fig. 3.3.

(Membership as of May, 1970; Association offices are in Washington, D.C.)

ABC-CLIO, Inc.	McGraw-Hill, Inc.
Arcata National Corporation	Microform Data Systems, Inc.
Aspen Systems Corporation	Microform Publishing Corporation
Auerbach Info, Inc.	Micromation Systems, Inc.
CBS/Holt Group	National Business Services, Micrographix
CCM Information Corporation	Division
Cahner's Publishing Company, Inc.	National Cash Register
Cognitive Systems, Inc.	New York Times
Computerpix Corporation	Octal Systems, Inc.
Congressional Information Service	Panta, Inc.
D.A.T.A.	Pergamon Press, Inc.
Eastman Kodak Company	Plenum Publishing Corporation
Herner & Company	Pharmaco-Medical Documentation
International Business Machines	Predicasts, Inc.
Corporation	Profile Communications, Inc.
Infodata Systems, Inc.	Quantum Science Corporation
Information Company of America	Showcase Corporation
Information Design, Inc.	Specialized Business Services, Inc.
Information Retrieval, Ltd.	Stroud, Bridgeman Press, Ltd.
Information Systems Corporation	Technical Information, Inc.
Institute for Scientific Information	Time, Inc.
International Data Corporation	Westinghouse Telecomputer Services
International Systems Design, Inc.	Corporation
The Lawyers Co-Operative Publishing	John Wiley & Sons
Company	The H. W. Wilson Company
Leasco Systems and Research Corporation	Xerox Corporation

Fig. 3.3.   Information Industry Association Corporate Membership.

3.14. CONTINUING EDUCATION

It is important for the abstractor to keep his "tools" sharp and workable and to keep himself well informed about new developments that are likely to affect his work. One obvious area in which it is important to keep up to date is the subject field of the abstractor (e.g., polymer chemistry, avionics, genetics). This is done almost automatically in the course of abstracting in the field of specialization.

The abstractor can follow important developments in other fields of science by reading multidisciplinary publications such as *Science, Scientific*

American, and *Nature.* Developments in the information sciences also bear careful watching. Some of the important sources in this field are:

1. *Journal of the American Society for Information Science* (Washington).
2. *Journal of Chemical Documentation* (American Chemical Society, Washington).
3. *Information Science Abstracts* (Yale University, New Haven, Connecticut).
4. *Annual Review of Information Science & Technology.**
5. *News from Science Abstracting & Indexing Services* (NFSAIS, Philadelphia).
6. *Scientific Information Notes* (Science Associates/International, New York).

An interesting observation is that abstractors are, themselves, users of abstracts in their continuing education efforts.

3.15. ABSTRACTING AS A CAREER

Few people abstract full time—even abstractors who work for regularly published major abstracting services may have other responsibilities (e.g., indexing) which provide job variety.

Abstractors who work for large abstracting services can aspire to challenging positions of substantial responsibility and diversity. They can rise in the ranks to hold such titles as section editor, group leader, associate editor, assistant editor, senior associate editor, senior staff advisor, computer index coordinator, subject heading specialist, production manager, assistant managing editor, managing editor, assistant director, and director.

At the upper levels of a large abstracting organization, abstractors can function much like managers in any other large organization, and they can expect the same kinds of responsibilities and rewards for their efforts.

In addition, much abstracting is done by information scientists who work in industrial and other information centers. Their activities are a mixture of abstracting, translating, indexing, literature searching, technical editing and writing, and other related activities. They are expected to be all-around performers, with opportunities, challenges, and rewards to match.

* Especially noteworthy is "Abstracting and Indexing Services in Science and Technology," by Stella Keenan, Chapter 9, pp. 273–303, in Vol. 4 (1969), Encyclopedia Britannica, Chicago, Carlos A. Cuadra, editor.

The diversity of abstracting and related activities, especially in industrial organizations, makes the career attractive to many. These kinds of activities bring the abstractor into contact with scientific, management, and legal personnel at all levels. In addition, the opportunity to keep up to date with important developments in many fields and to become recognized as a highly skilled communicator is an attractive feature of the career.

chapter 4
Selecting Material for Abstracting

Before beginning to write his abstract, the abstractor should ask this question: Is the original document worth abstracting in the first place? This may seem to be an obvious question, but it is one that needs answering before the abstracting process is started. The abstractor must decide for himself, or he must be advised by the editor of the abstract journal. Also, there are usually some inherent boundaries or outer limits based on the availability of resources to the abstracting service.

The abstractor should take into account the characteristics of the literature under consideration for abstracting. He knows, for example, that much of what is published is not significant because individuals in many organizations are under pressure to publish if they are to further their careers. To put this another way, not all that is published is equally significant and not all is worth abstracting. The extent to which this and other considerations affect selection of material to be abstracted is in turn influenced by the nature and resources of the abstract bulletin, and by the needs of its clientele. It is worth mentioning again that a major factor is the availability of time and funds; abstractors who have plenty of both can afford to include more documents than those less fortunate.

The abstractor may want to be guided by a tabulation such as the following in making his selection of documents to be abstracted:

Consider *in favor of* abstracting documents which are:	Consider *not* abstracting documents which are:
1. Pertinent to the interests of the clients of the abstracting service.	1. Not pertinent or of marginal interest only.
2. Novel (original) contributions.	2. Repetitious (nonoriginal) material.
3. Final reports and other reports adequately supported by data or other evidence.	3. Preliminary progress reports inadequately supported by data or other evidence.
4. More difficultly accessible publications; foreign documents.	4. Readily accessible, widely circulated domestic articles (perhaps just cite these or abstract very briefly).
5. Significant material.	5. Trivial material.
6. Published in reputable sources.	6. Published in sources of unknown or questionable reputation.
7. Published in sources an abstracting service has committed itself to cover by policy.	7. Published in sources usually excluded by policy.

Some persons think that only the user can decide what is worthwhile. If carried to extremes, this belief could mean that "everything" must be abstracted—a policy which, in our opinion, is not practical.

Lancaster (12) has discussed use-based selection policies from the viewpoint of the indexer; his comments are also of value to the abstractor. He suggests that a study be made of use of the service or system. Selection policies can then be based on the frequency with which documents from various sources or languages are requested and found of value. The implication is that those sources (e.g., journals) and languages most frequently requested or used are abstracted in preference to those less frequently in demand. Lancaster notes that "if our collection consists of well-defined document series . . . , we can improve our economic efficiency by investing most of our . . . resources on series most likely to produce the greatest payoff in terms of input cost over volume of use." The techniques suggested by Lancaster are most likely to be successful when the users are both available and willing to yield the kind of "feedback" desired.

Once selection policies have been decided on, a concerted effort is needed to make sure that the kinds of materials to be abstracted are received as promptly and completely as possible. In larger abstracting operations this function (acquisition) requires a skilled, full-time staff, and a constant vigil is needed to identify and acquire new materials of possible interest.

Selection policies should be reviewed at least every six months to ensure that they are up to date. Abstractors should take care to inform *abstract users* of details of selection policies so that users will know what to expect.

The following sections discuss a few special cases in selection policies.

4.1. GOVERNMENT-SPONSORED TECHNICAL REPORTS

These reports (usually published as soft-covered "separates") pose a special problem to the abstractor. There are so many thousands of them, and a substantial proportion are classified as secret or confidential. These reports are not usually subjected to outside editorial and review processes, but they may represent very important work reported for the first time.

Some government technical reports are admittedly ephemeral progress reports that represent only partial, inconclusive work often rapidly superseded by more definitive work. Other so-called technical reports are really concerned with administrative matters only, with minimum technical content. These rarely deserve anything more than a citation, if that, and then only to help "follow the trail" of an elusive project.

Most government reports are abstracted promptly in highly efficient, government-sponsored abstracting services. Most of the more significant unclassified advances are eventually published in conventional journal literature; therefore, abstracting both reports and the corresponding literature in one abstracting service means duplication of effort. But some of the major externally published nongovernment services will abstract unclassified government reports if these reports are deemed sufficiently important.

A reasonable approach to the abstracting of unclassified technical reports (for nongovernment abstracting bulletins) is to abstract only those government reports that (a) cover significant technical work, (b) include complete and well-supported data, and (c) are pertinent to the interests of audience being served.

4.2. BOOKS

Some books are really collections of papers in bound form. These include, for example, the proceedings of technical meetings and symposia. Each contribution to such a book deserves a separate abstract. For convenience, the abstracts are grouped so that the citation to the original need not be repeated.

Most other kinds of books, unless they contain original, previously unpublished work or literature review material, are usually not abstracted. But relevant books can be *cited* when appropriate. The title is usually sufficient to indicate content. If it is not, it should be expanded by the abstractor sufficiently to make the meaning clear.

For important books, it may be advisable to list the titles of individual chapters, particularly when there are contributions by a number of relatively prominent authors.

4.3. REVIEW ARTICLES

Although review articles usually require much time and skill to prepare, the abstractor should give them relatively brief treatment. The abstractor should note the number of references covered and scope of the review. But if a review article is significant, it may well be worth a full abstract. Examples of significant review articles include the first review of a new or rapidly expanding technology, or a review of a controversial subject. The abstractor may want to start his abstract with a phrase such as:

This paper brings together for the first time widely scattered information on . . . .

4.4. EFFECT OF DOCUMENT CONTENT; SOME FACTORS

Assuming that the abstractor considers a document worth abstracting, he must also decide what parts of the document are worth abstracting. The content of the document has a bearing on this. Some of the factors the abstractor will want to consider are:

Negative results
Safety, toxicology, and related matters
Business aspects
Recommendations for future work
Experimental details

4.4.1. Negative Results

Should negative results be abstracted? For purposes of this book, we can define negative results as those meeting a very simple criterion: *some* or *all* of the desired results were not achieved by the scientist in his work. If the scientist reports such work in the original document, as he should, the abstractor should include the negative results reported.

The abstracting of negative results is important and worth emphasizing.

If something does not go as expected, other investigators may want to see just why the experiment did not work and whether the experiment will work under different conditions. Negative results can lead to new, unexpected and highly positive work. In addition, negative results can be symptomatic of hazardous conditions about which other investigators should know before they attempt to repeat the experiment.

4.4.2.   Safety, Toxicology, and Related Matters

Information about the safety aspects of a piece of research or other technical investigation should be abstracted as fully as possible if reported in the original document. By so doing, the abstractor alerts future investigators to potential hazards such as fire, explosion, and toxicity. The abstractor can help save lives.

4.4.3.   Business Aspects

The business or economic aspects of science and technology are especially important to profit-making organizations, and many investigators in other kinds of organizations are equally interested in these aspects. Managers and engineers who scale up laboratory (or pilot plant) studies to full (or plant) dimensions are especially cost conscious. Also, many technical managers are interested in market-research data.

The abstractor can help management to arrive at business decisions by covering parts of documents that relate to the business or economic aspects of science and technology. The accompanying data can be included in abstracts. Specific production cost estimates are useful as are market and other data relating to the economic potential of a product or process.

There is nothing wrong with including dollar signs or other monetary designations and data in abstracts of technical documents that contain costs or other economic data. To the contrary, by inclusion of significant economic data in his work, the abstractor shows that he is aware of the realities of this world. For example, an abstract such as the following is perfectly acceptable even in an abstracting bulletin which emphasizes the scientific aspects of a field:

New 123 process for mfg. of product XYZ will be used by ABC Co. in plant to be built in Main Town. Plant will cost $8 million and will have annual capacity of 5 million lbs. when construction is completed in 197x. The 123 process is covered by U.S. Patent x,xxx,xxx (previously abstracted, see p. 842, 197x). Market potential for use of Product XYZ in steel industry is estimated at 10 million lbs./yr.

4.4.4. Recommendations for Future Work

Most scientific papers include recommendations for future work—recommendations often based on speculation about the results of present work. Should the abstractor cover such material in his abstract?

We think that the answer to this question will ordinarily be "no." The main reason is that the need for conciseness makes it difficult enough to include all findings fully substantiated by experimental results, much less to include speculative material. The author of the original document is fully entitled to include such material, but the abstractor is well advised to limit himself to conclusions supported by evidence.

4.4.5. Experimental Details

Many experimental details—especially information on physical apparatus or laboratory setups—should not ordinarily be abstracted because of space limitations. However, there are some exceptions. For example:

1. A paper that is primarily concerned with developing a new experimental technique will be abstracted with full detail about technique.

2. Basic conditions (e.g., temperature, pressure, and other basic data) should be included in the abstract.

3. A paper concerned with developing data regarding physical constants should include details about apparatus and techniques.

chapter 5

First Step: Citing
the Reference

The abstractor should make sure that he has on hand a fully legible and complete copy of the document to be abstracted. He can then proceed to write, dictate, or type the citation according to some very definite steps, as outlined below. Since abstracts are secondary sources, the reference is of special importance and deserves careful treatment and attention.

Citing the complete reference accurately and concisely is a cardinal principle of abstracting. The principle is so important that the various parts of the reference are worth spelling out *although the form and the exact sequence in which these various parts may appear is subject to the guidelines of the organization (or those which the abstractor may set down for himself)*. We recommend the following sequence of the basic "data elements":

5.1. FULL TITLE OF THE DOCUMENT

This ordinarily is the first part of any abstract reference. For foreign language documents, the abstractor should translate the title of the original document into English. The original foreign language title need not appear in the citation—just the English translation. But some abstract journal editors prefer to print the title both in the original language *and* in English in the interests of scrupulous accuracy. Still other editors print the title in the original language only. It is a matter of preference, which varies with the editor.

In all cases, the abstractor should take a careful look at the title of the original document to determine whether it is meaningful and specific. If it is not, the abstractor can make it meaningful by expanding or otherwise clarifying the meaning. Brackets can be used to indicate words added by the abstractor and to ensure proper identification of the document title by the reader. For example:

Original title: The Supersonic Transport (This title is too general.)

Abstractor's revised title: [Design of Proposed Alternates for] The Supersonic Transport [Aircraft Engines]

Abstractors find that title expansion or clarification is especially needed when working with patents. Official patent titles are often intentionally brief or cryptic. For example, a patent title such as

Hydrazine

can be expanded to something more meaningful and specific:

[Manufacture of] Hydrazine [by Process XYZ]

Here is another example of title improvement:

Original title: Process

Better: Process [for Manufacture of XYZ]

Still better: Process [for Manufacture of XYZ by Method ABC]

The importance of a good title deserves special emphasis. Some readers get no further into the abstract than the title. If titles do not indicate something of interest, they may read no further. For example, a reader may be interested only in *uses* of hydrazine. The one word title, "Hydrazine," does not help this reader very much, if at all, especially if the document deals only with hydrazine *manufacture*. If titles are cryptic, readers are forced to plow through what may be lengthy abstracts—all in vain if the general subject is not of interest. *Abstractors can be kind to their readers by providing unambiguous titles.*

Abstractors should clearly identify titles of articles that may be part of a series. For example:

Manufacture of New Steels. Part XIX: High Strength Steels for Aerospace Applications

Note the use of the title of the series as well as the title of the individual *part* of the series.

5.2. AUTHORS

Decisions on how to cite authors affect the construction of author indexes to abstracts and should be coordinated with the indexer.

Our suggestions for citing authors are as follows: Give the last name(s) first, and then full given name(s)—or initial(s) if full given names are not supplied—exactly as they are printed in the text of the original document. For example:

> Smith, J. B., Jr.; Brown, Thomas B.; Jones, W. Roger

But note that the exact form of the citation would depend on the guidelines that are used. For example, either "W. Roger Jones" or "Jones, W. Roger" could be specified as the preferred form.

The citing of initials only for given names might lead to difficulty in location of the original, but saves time and space. For this reason it is usually an acceptable practice in smaller, in-house operations, where time is at a premium.

Special care should be taken in transliteration of names into English from such languages as Russian. The abstractor will find information on this point in *Directions for Abstractors,* published by Chemical Abstracts Service.

5.3. AFFILIATION AND LOCATION OF AUTHOR

The organization and its location is cited immediately after the names of the authors:

> Smith, X. Y. (University of Michigan, Ann Arbor)
> Jones, Alfred B. (Research Center, Olin Corp., New Haven, Conn.)

The geographic designation is important since many organizations have more than one location. Provision of full location data by the abstractor enables the reader (*a*) to make preliminary decisions about the pertinence of the original document to his work (for example, work done at a location which is known to be a center for a specific field of research of interest to the reader is likely to attract immediate attention) and (*b*) to contact the author of the original, if desired, to ask for reprints or for other reasons.

Even for in-house reports, organization (department) and geographic designation can be helpful and should be cited. For example:

> F. R. Jones (Chemicals Division, Niagara Falls, N. Y.)

In abstracts of patents, the name of the inventor is usually followed by the name of the organization to which the patent rights are assigned. Location is, by convention, usually not cited. For example:

Black, O. Y. (Olin Corp.)

Again, the information about the organization is important since it can help to identify immediate interest in a patent. For example, the reader may be interested only in patents assigned to a specific organization.

5.4. CITING A JOURNAL

The abstractor should abbreviate the journal title in accordance with one of the standard guides such as:

1. *American Standard for Periodical Title Abbreviations,* USASI (formerly ASA) Z39.5 (1963), United States of America Standards Institute, New York, N.Y.

A 1969 revision of this is now available. The United States of America Standards Institute is now known as the American National Standards Institute; standards are revised as required.

2. "Revised & Enlarged Word Abbreviation List for USASI Z39.5 (1963): American Standard for Periodical Title Abbreviations," National Clearinghouse for Periodical Title Word Abbreviations, Chemical Abstracts Service, Columbus, Ohio (see also supplements).

3. *Access,* Chemical Abstracts Service, Columbus, Ohio, 1969 (and supplements). This gives abbreviations (based on the standards cited above) for more than 30,000 titles and also has much valuable identification and library-holdings data.*

The abbreviated journal title should be followed by the volume number, inclusive pages, and year. If the page numbering begins anew with each issue, the issue or issue number must always be identified. Even if the page numbering does not begin anew, it is excellent policy to include the issue number for foolproof identification purposes. For example, title, author, and author's location are followed by a citation such as:

J. Amer.Chem.Soc. 89(26) 6807–6813 (1967)

The designation (26) indicates that this is issue 26 for the year.

In smaller, in-house organizations where time and space are important, briefer citations may be used, especially for well-known publications:

JACS **89** 6807 (1967) can be used in any chemical laboratory since every chemist knows that JACS = *Journal of the American Chemical Society,* and since this publication is widely available.

* Note: *Access* is now titled *CASSI* (Chemical Abstracts Service Source Index).

5.5. CITING A PATENT

The suggested sequence for citing patents is: title; inventor(s); organization to which patent is assigned; abbreviated name of issuing country; patent number; and date of issue. For example:

> Preparation of Stearoyl Chloride. Rothman, Edward S., Serota, Samuel (U.S. Department of Agriculture as represented by the Secretary of Agriculture), U.S. 3,497,536, 24 February 1970.

There are additional (mostly legal) details that the abstractor can include in patent citations, but a discussion of these is beyond the scope of this book. Consult references 9 and 10 cited at the end of this book for more information.

5.6. CITING A BOOK

A suggested sequence for book citation is: title; author(s); publisher; place of publication; date; number of pages; and price. If the title is in a language other than English, the English title may be given first, followed by the language of the original. For example:

> Medium Chain Triglycerides. Senior, John R.; Van Itallie, Theodore B.; Greenberger, Norton J.; Editors (University of Pennsylvania Press, Philadelphia, Pa.). 1968. 300 pp. $12.50.

5.7. NUMBER OF REFERENCES

Any document that has many references is usually of special value in following the trail of previous related research. Therefore, the abstractor sees to it that documents that contain many references are so specified by indicating the number of references included in the document. This information is best given at the end of the abstract, thus: 27 ref., or 19 references. Similarly, if an article (especially a review article) has no references, this should be specified in so many words, thus: no references.

5.8. LANGUAGE OF THE ORIGINAL

The importance of specifying language in the abstract citation is that many scientists will look at the full originals of only those documents written in languages that they can read—providing, of course, that the subject matter is of sufficient interest. Also, many scientists fail to appreciate that certain journals with foreign language titles may contain articles in English, and that the reverse is also true: journals with English language titles can contain articles written in languages other than English. In fact, we now live in a world so multilingual that some abstractors think it is good practice to *always* make note of the language unless it is very obvious.

For example, an article in *Angewandte Chemie (International Edition in English)* is obviously in English, and a further note to this effect in the citation is not essential. But a German language article in the British publication *Tetrahedron Letters* should be so specified, as should an English language article in the Swedish journal *Svensk Papperstidning*. Another option is to cite languages only for those documents in languages other than English if this policy is made clear to the user.

Notations of language are best made at the end of the citation, immediately before the abstract proper. For example:

(in English) or (Eng.)
(in German) or (Ger.)
(English translation)
(Russ.)
(English summary)
(English summary of machine translation)

5.9. MORE ABOUT CITATIONS

The purposes of all this attention to detail in the citation are: to give as much identification as possible so that the full original document can be readily obtained if desired; to fully identify the author and his location; and to enable the reader to decide whether to look at the original document (although the abstract proper is more useful to the reader in achieving this objective).

There are obviously other kinds of documents than those for which we have suggested citations. The same general principles hold true for these kinds of documents as well.

The various parts of the citation (and the rest of the abstract) are shown diagrammatically in Fig. 5.1.

Although abstractors should fully identify original documents, *the sequence of the items in the citation is not critical.* The important thing is that all the information be there in clear and unmistakable form. However, some organizations have guidelines, as previously mentioned, and if the abstractor works for, or contributes abstracts to, one of these organizations, it is clearly advisable to follow the prescribed style of citation. Also, if the citation is input to a computer, it may be necessary to follow a prescribed sequence. Hence, it bears repeating that we have only *suggested* or *recom-*

mended; in so doing, we have at least alerted the abstractor to the kinds of decisions he must make in the course of his work with citations.

With the exception of the clarification of the title of an original document to make it more meaningful, an intelligent, well-trained clerk can often prepare the citation for checking by the abstractor. This kind of assistance can save the abstractor considerable time.

The abstracts

The information contained in the abstracts entries is described in the following examples:

Abstract entry from a paper appearing in a periodical journal

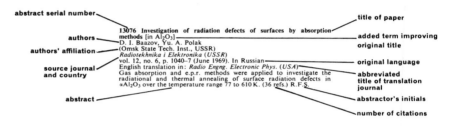

Abstract entry from a patent

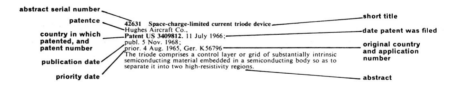

Abstract entry from a technical report

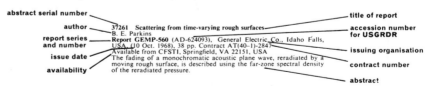

Fig. 5.1. Parts of the citation and other parts of the abstract entry. *Source. Electrical and Electronics Abstracts,* (January, 1970, page B 2) published jointly by the Institution of Electrical Engineers, and the Institute of Electrical and Electronics Engineers, Inc. Reprinted with permission.

chapter 6
Kinds of Abstracts

Abstractors make distinctions between two major kinds of abstracts—*indicative* and *informative*—and can also identify several other kinds of abstracts. *Indicative* abstracts merely tell briefly what the document is about. *Informative* abstracts are longer and present the essential data and conclusions so that the reader has no need to refer to the original document. However, we are not implying that an informative abstract is a substitute for the original document.

In practice, the distinction between indicative and informative abstracts blurs. Abstracts have evolved from rigid formats into flexible, viable modes of expression, highly responsive to user needs—and this is exactly what the good abstractor tries to achieve. There is nothing wrong with mixing indicative and informative abstracts, and the gray areas in between, in the same group of abstracts, as long as user needs are served.

6.1. INDICATIVE ABSTRACTS

Here is a fictitious example of an *indicative* abstract:

> The finishing of textiles by process ABC to achieve water repellency is considered superior to finishing by process XYZ. Factors considered include durability, appearance, cost, and speed.

This kind of abstract tells what the article is about. The advantages are that it can be written quickly and economically. It will probably satisfy an audience of generalists. But for the scientist specifically interested in the subject (say, a textile finishing specialist), it is probably not satisfactory because of the lack of detail.

6.2. INFORMATIVE ABSTRACTS

The indicative abstract shown above can be rewritten in the form of an *informative* abstract:

> ABC is judged a more satisfactory process for the waterproofing of synthetic textiles than XYZ. The ABC process yields a product of 20 percent greater durability as judged by standard test #1234. It also yields a better appearing product based on the votes of a panel of 25 textile finishing specialists. The cost of ABC is claimed to be 10 per cent less per square yard although specific cost data are not given. The two processes are about equal in processing speed.

A further comparison between indicative and informative abstracts is offered by the sample abstracts shown in Figs. 6.1 and 6.2 published in *Metals Abstracts* and in its predecessor publication, *Review of Metal Literature*. The abstracts in *Metals Abstracts* are detailed and informative, whereas those in *Review of Metal Literature* in its earlier years were indicative. The change from indicative to informative abstracts in the *Review of Metal Literature* (previously published by the American Society for Metals) was made in 1965, although the merger of that publication with *Metallurgical Abstracts* (previously published by the Institute of Metals) was not made until the beginning of 1968. *Metals Abstracts* is the successor publication, published jointly by the American Society for Metals and the Institute of Metals (London); this is an excellent example of international cooperation in abstracting.

6.3. EXTRACTS

Although quite a bit of detail is presented in informative abstracts, some readers may require still more detail, such as tables of comparative data and other excerpts from the original. For example, the material presented in the Copper Development Association publication, *Extracts of Documents on Copper Technology,* consists of lengthy extracts. If the quantity of documents involved is large this approach can be expensive, but the convenience to the user may outweigh everything else.

77-R. (German) STRESS CORROSION OF AlZnMg 3 AS AFFECTED BY TEMPERATURE. Wolfgang Guhl. Zeitschrift fur Metallkunde, v 53, no 10, Oct. 1962, p 670-675.

Corrosion testing of tensile stressed AlZnMg 3 alloy strip specimens containing 4.94% Zn, 3.02% Mg, 0.06% Cu, 0.34% Mn, 0.33% Si, 0.32% Fe, 0.14% Cr, 0.017% Ti and 0.00036% Be in 3% NaCl solution at 20 to 70° C. Corrosion rate as affected by stress and temperature. Interpretation in terms of vacancy migration. 19 ref. (R1d, 2-61, 3-66; Al-b, Zn, Mg, 4-53)

78-R. CORROSION STUDY OF STEEL PILING IN SERVICE. Technical News Bulletin, National Bureau of Standards, v 46, no 11, Nov. 1962, p 164-165.

Investigation of the dependence of piling corrosion on soil conductivity, composition, acidity and oxygen content as well as on size and length of service. (R-general, 2-66, 3-67;ST)

79-R. A BASIS FOR DESIGN OF ALUMINUM ALLOYS FOR HIGH TEMPERATURE WATER SERVICE. R. L. Dillon and H. C. Bowen. Corrosion, v 18, no 11, Nov. 1962, p 406t-416t.

Investigation of the corrosion rate dependence on exposure time, heat treatment, alloy composition and constituent purity as well as on casting variables for Al alloys, including X-8001, A-203X and a 1.8% Fe-1.2% Ni-0.002% Si alloy. 12 ref. (R-general, 2-64, 2-60, 3-67; Al-b)

80-R. CORROSION OF STEEL IN CONCENTRATED LITHIUM HYDROXIDE SOLUTION AT 316° C. M. C. Bloom, W. A. Fraser and M. Krulfeld. Corrosion, v 18, no 11, Nov. 1962, p 401t-404t.

Dependence of the corrosion rate, pitting and surface microstructure of mild steel on solution concentration during corrosion testing using a capsule containing the specimen and corrosive agent. Lithium hydroxide is found to produce much reduced corrosion as compared with a similar sodium hydroxide solution. 20 ref. (R6j; ST)

81-R. CORROSION OF ZINC PRESSURIZED DILUTE AQUEOUS SOLUTION

SYSTEMS. R. C. Weast and S. L. Shulman. Corrosion, v 18, no 11, Nov. 1962, p 417t-420t.

Samples of high-purity Zn are corrosion tested in tap water and water with high ion concentrations in the 50 to 80° C. range as a function of pressure. Reaction vessels and testing procedures. 10 ref. (R-general, 2-61, 2-66, 3-74; Zn)

82-R. (German) CONDENSOR TUBE CORROSION AND MATERIALS. PT. 1. BRASS. F. W. Nothing. Metall, v 16, no 11, Nov. 1962, p 1089-1096.

Microscopic and macroscopic analysis of various types of corrosion fatigue, pitting, intercrystalline corrosion and stress corrosion on brass condensor tubes with varying As content. 20 ref. (R-general, M21, M28g; Cu-n)

83-R. (French) METHODS OF CONTROL OF METALLIC COATINGS IN RELATION TO CORROSION PROTECTION. P. Morisset. Corrosion et Anticorrosion, v 10, no 9, Sept. 1962, p 281-285.

Determination of influence of adherence, thickness and structural uniformity of coating on corrosion resistance of Zn or Ni plated iron or steel components using atmospheric corrosion tests, or the TIM test using sulfur gas, saline solution and varying amounts of humidity. (R3, 1-54, L17b, Q10c; St, Ni, Zn)

84-R. (French) ECONOMIC ASPECTS OF CORROSION PREVENTION IN REFINERY RESEVOIRS. P. W. Sherwood. Corrosion et Anticorrosion, v 10, no 9, Sept. 1962, p 286-291.

Corrosion resistance of Al, steel and forged iron sheets painted or cathodic protected by Mg, Zn, Al, C, Fe or Pt anodes. 5 ref. (R-general, R10d; Al, Fe-b, ST)

85-R. (French) STUDY OF THE FORMATION OF RED FUMES BY IMPACT OF AN OXYGEN JET AT THE SURFACE OF A LIQUID METAL. G. Urbain. Revue de Metallurgie, v 59, no 9, Sept. 1962, p 725-732.

Influence of refining temperatures from 1710 to 2310° C. and alloy content on presence of red fumes during oxidation and iron vaporization of Fe, Fe-C, Fe-C-Si and Fe-Si alloys. 21 ref. (R1h, 2-62; Fe-b, Si, C)

Fig. 6.1. Examples of indicative abstracts. *Source. Review of Metal Literature*, 1963. Reprinted with permission.

9-C. PRECISION MELTING AND CASTING.
A. Dunlop. Foundry Trade Journal, v 113,
no 2395, Nov. 1, 1962, p 541-549.
Melting, casting and grain refining
of Al, Cu and Ni alloys. Determination of grain size, density and creep-
rupture properties. 7 ref. (C5, E25g;
Al-b, Cu-b, Ni-b)

10-C. PREPARATION OF ZIRCONIA SOLID
SOLUTIONS BY COPRECIPITATION. A. J.
Hathaway and Abraham Clearfield. (Abstract
of paper presented at meetings of the Ceramic
Association, 1961.) American Ceramic Society,
Bulletin, v 41, no 9, Sept. 1962, p 583.
Comparison of the sinterability of
$ZrO_2-Y_2O_3$, ZrO_2-CaO and ZrO_2-CeO_2
after coprecipitation and calcining.
Sintering of binary coprecipitates from
200 to 3000° F. (C27, H15, 2-62;
Zr-b, Ca, Ce, Y)

11-C. (Translation-PrimSources) REFINING
ZINC BY RECTIFICATION. V.V. Krapukhin,
L. G. Povedskaya and S.A. Ershova. Soviet
Journal of Nonferrous Metals, v 2 no 6, June
1961, p 24-28.
High-purity Zn containing almost no
Cd or Pb is obtained by rectification using
a column with pure graphite plates distri-
buted in zig-zag patterns along the column
height. 8 ref. (C22h;Zn)

12-C. (Translation-PrimSources) REMELTING
HIGH-PURITY CATHODIC ZINC. D. P.
Zosimovich and N.A. Shvab. Soviet Journal of
Nonferrous Metals, v 2, no 6, June 1961, p 29-33.
Remelting and casting of high-purity Zn
without contamination using a smelter made
up of a graphite crucible in a hermetically
sealed shell and a casting chamber of graphite
molds controlled by thyratrons. 8 ref. (C5;
Zn)

13-C. (German) KINETICS OF THE
CHLORINATING EVAPORATION OF ZINC
FROM ZINC OXIDE USING CALCIUM
CHLORIDE. Ulrich Kuxmann. Zeitschrift
fur Erzbergbau und Metallhuttenwesen, v
14, no 6, Sept. 1962, p 447-462.
Vapor pressure measurements and
investigation of the kinetics of the
chlorinating evaporation of Zn from
ZnO from 850 to 1050° C. Effect of the
kinetics of the process on extraction

data including CaO content of melt,
$CaCl_2/ZnO$ ratio and carrier gas velocity.
27 ref. (C26, P12c, 2-62; Zn)

14-C. (Russian) REDISTRIBUTION
OF CARBON AND TUNGSTEN IN COLUM-
BIUM DURING ZONE REFINING. A. I.
Evstiukhin, V. V. Nikishanov and I. V.
Milov. Izvestiya Akademii Nauk SSSR--
Metallurgiya i Toplivo, no 3, 1962, p 98-
101.
Zone refining of Cb, containing C
and W as impurities in an electric arc
furnace, followed by evaporation and
reduction. Determination of W and C
contents using C^{14} and W^{182} isotopes.
Effect of zone refining on mechanical
and physical properties. 6 ref. (C28k,
P-general, Q-general, 1-59, 2-60; Cb,
C, W)

15-C. (Translation-Prim Sources)
PROCESSING WASTE COPPER SLAGS.
A. S. Pen'ko. Soviet Journal of Nonferrous Metals, v 2, no 8, Aug. 1961,
p 100-101.
Smelting and reducing processes
to recover Fe and Zn from waste Cu
slags. (C21; Cu, RM-g)

16-C. ELECTRIC SMELTING OF ANTIMONY
DUST. I.I. Kershansky and L.N. Rogova.
Tsvetnye Metally, no 6, 1962, p 39-41.
Granulation and smelting of Sb dust
(42.14% Sb, balance Pb, As, Fe, Zn, Na, S,
Se, Tl, Au and Ag) in magnesite-lined
electric furnace, using Na (10-20%) as flux
and coal (6%) of given granularity as reducing agent. Quantitative determination of the
smelting product components: black Sb,
slag-matte melt and vapors. Comparison of
reducing efficiencies of electric furnaces with
those of reverberatory furnaces. (C21d, 1-52,
2-60; Sb)

17-C. (German) PROGRESS IN THE
PRODUCTION OF TITANIUM WITH PAR-
TICULAR REFERENCE TO GERMAN
PATENTS. K. Bayer. Metall, v 16,
no 10, Oct. 1962, p 975-978.
Review of extraction and refining of
Ti metal, including reduction of $TiCl_4$
with Mg or other alkaline earth metals,
and Na and other alkaline metals; thermal decomposition of titanium halides;
reduction of TiO_2 by C, Na and Ca and
other reducing agents; refining of Ti
metal and scrap, including fused salt
electrolysis. 75 ref. (C-general, 10-54;
Ti)

Fig. 6.1. *(Continued)*

61-C. HIGH-PRODUCTION CASTING OF VACUUM-MELTED SUPERALLOYS. Machinery, v 69, no 4, Dec. 1962, p 152.

Operation and design of semicontinuous induction furnace unit which melts, investment casts and cools superalloys under a vacuum. Fired-ceramic molds made by both the frozen mercury and lost wax investment processes are preheated from 1700 to 1900° F. for 8 hr before pouring. (C5, 1-52, C25; SGA-h)

62-C. (Russian) NEW BEARING ALLOYS. N. P. Nikolaichik. Liteinoe Proizvodstvo, no 11, 1962, p 37-38.

Shrinkage, yield strength, elongation, resilience and hardness as affected by composition of sleeve bearings cast from a new alkusip alloy composed of Al, Cu, Fe, Pb and Si. Melting, mixing, alloying and casting techniques. (C5, P10c, Q-general, T7d, 2-60; Al-b, Cu, Fe, Pb, Si)

63-C. A SURVEY OF SOME OF THE NEW METALS AVAILABLE TO THE ENGINEER. N. P. Inglis. Institute of Metals, Journal, v 91, no 4, Dec. 1962, p 121-133.

Procedures for vacuum melting, electron beam melting, hot working, cold rolling, cold pressing, drawing, extrusion and welding of Ti, Zr, Hf, Be, Cb and Ta. 20 ref. (C25, C5k, F-general, 1-66, 1-67, K-general; Be, Cb, Hf, Ta, Ti, Zr)

64-C. (Translation-PrimSources) INDUSTRIAL TESTS OF A NEW CONTINUOUS DECOPPERING METHOD FOR CRUDE LEAD. Ya. Z. Malkin, M. P. Smirnov, V. Ya. Sergienko, G. I. Kozhevnikova, E. I. Kalnin and N. G. Tarkhov. Soviet Journal of Nonferrous Metals, v 2, no 3, Mar. 1961, p 15-22.

Previously abstracted from original. See item 282-C, 1961. (C21e, 2-61; Pb, Cu)

65-C. (Translation-PrimSources) STARTING A 32,000-KVA THERMIC ORE SMELTING FURNACE. M. D. Sudarev, N. A. Komnatny, E. V. Berdennikov and N. C. Sobolev. Soviet Journal of Nonferrous Metals, v 2, no 3, Mar. 1961, p 23-31.

Previously abstracted from original. See item 283-C, 1961. (C21d, 1-52; Cu, Ni)

66-C. (Translation-PrimSources) SHAFT FURNACE SMELTING OF COPPER USING AN OXYGEN ENRICHED BLAST. K. I. Artamonov, N. I. Lebedev, E. E. Ergaliev, A. K. Lesechko, M. V. Yakushin, V. N. Kazakov, N. G. Bryukhanov, L. N. Nikitina and F. I. Khvesyuk. Soviet Journal of Nonferrous Metals, v 2, no 3, Mar. 1961, p 32-39.

Previously abstracted from original. See item 284-C, 1961. (C21a, 1-52; Cu)

67-C. SEPARATION OF HAFNIUM FROM ZIRCONIUM AND THEIR DETERMINATION: SEPARATION BY ANION-EXCHANGE. Lawrence A. Machlan and John L. Hague. Journal of Research-National Bureau of Standards-Sect. A: Physics and Chemistry, v 66, no 6, Nov-Dec. 1962, p 517-520.

Separation by elution techniques of Hf from mixtures of Hf and Zr containing 80% or less Zr. Determination is accomplished by precipitation, ignition and weighing the materials as HfO_2 and ZrO_2. Curves show elution behavior of the elements. 9 ref. (C27; Zr-b, Hf)

68-C. (French) METAL ORES. Pro-Metal, v 15, no 89, Oct. 1962, p 719-721.

Review of dry and wet extraction techniques for ores containing Cu, Zn, Ni and Pb. (C-general, 2-66; Cu, Ni, Pb, Zn)

69-C. (German) METAL EXTRACTION BY LEACHING AND PRECIPITATION UNDER PRESSURE. Pt. 1. J. Gerlach. Metall, v 16, no 12, Dec. 1962, p 1171-1179.

Thermodynamic and kinetic considerations in the pressure leaching and precipitation of metals. Examples of pressure leaching of oxides and sulfides in the extraction of Cu, Au, Ag, U, Co and Ni. 172 ref. (C19n, C27; Cu, Au, Ag, U, Co, Ni)

70-C. GASES IN COPPER BASE ALLOYS. J. F. Wallace and R. J. Kissling. Foundry, v 90, no 12, Dec. 1962, p 36-39.

Effects of hydrogen content and pouring temperatures from 1050 to 1250° C. on gas porosity during melting of Cu alloys in crucible furnaces; influence of alloying elements, such as Al, Sn and Ni, on hydrogen solubility. Methods for

Fig. 6.1. *(Continued)*

measuring the gas content of the melts. (C5, 2-60, 2-62; Cu-b)

71-C. (German) SELF-SUSTAINED ELECTRICAL DISCHARGE AS HEAT SOURCES IN MODERN VACUUM MELTING. P. Schadach. Elektrowarme, v 20, no 11, Nov. 1962, p 580-588.

Theoretical analysis and experimental investigation of the use of glow and arc discharges in vacuum as a mode of heating. Technical applications in vacuum melting and heat treating of refractory metals, including Ta, Cb, Mo, Zr, Ti and W. 13 ref. (C5, 1-73, J-general; EG-d37)

72-C. REDUCTION OF COPPER OXIDES BY MOLECULAR HYDROGEN. Y.N. Trehen. Zeitschrift fur Anorganische und Allgemeine Chemie, v 318, no 1-2, Sept. 1962, p 107-112.

Analysis of the growth of Cu films and polycrystalline Cu obtained by reduction of CuO, Cu_2O and mixtures of the two oxides with dry hydrogen at 300° C. X-ray structure determination for both oxides and reduced Cu. 21 ref. (C 26; Cu, 14-68)

73-C. ZONE REFINING. David Fishlock. Metalworking Production, v 106, no 51, Dec. 19, 1962, p 41-43.

Operating details for zone refining metals and semiconductors to obtain high purity, including special techniques for refractory and reactive metals. (C28k; EG-d37, EG-j)

74-C. A DISCUSSION ON THE REDUCTION OF SILICON DIOXIDE IN BLAST FURNACE. A.T. Pal. Institution of Engineers (India), Journal, v 42, no 11, Pt MM 3, July 1962, p 88-95.

Effects of temperature and slag basicity, composition and thickness on the reduction of SiO_2 with C, CO and CO_2 from 702 to 1557° C. in the blast furnace to produce ferrosilicon. (C21a, 2-62; RM-q, Si, 14-68)

75-C. (German) CONTRIBUTION TO THE VACUUM MELTING OF HEAT RESISTANT NI-CR-BASE ALLOYS. Rudolf Ehrt, Klaus Edelmann and Hans Stumpf. Neue Hutte, v 7, no 11, Nov. 1962, p 674-679.

Melting of heat resistant alloys containing 14.74-15.00% Cr, 1.8-2.3% Fe, 0.12-0.25% V, 0.23-0.34% Si, 4.79-5.10% W, 2.70-2.90% Mo, 1.87-2.59% Ti, 2.44-2.59% Al, 0.07-0.10% C and the rest Ni, in an open induction furnace and remelting in a vacuum arc, electron beam or vacuum induction furnace. Deoxidation, desulfurization, evaporation and reaction with furnace lining as influenced by the vacuum. 17 ref. (C5, 1-73; Ni, Cr, SGA-h)

76-C. (English) THE EXTRACTION OF MAGNESIA FROM SEA WATER. T.D. Ionescu and Al. Braniski. Revue Roumaine de Metallurgie, v 7, no 2, 1962, p 329-335.

Production of MgO by extracting Mg $(OH)_2$ from sea water using nonferrous metallurgical slag and HCl, followed by burning of $Mg(OH)_2$ at 1740° C. and subsequent sintering at 1600° C. 19 ref. (C19, H15, 2-62; Mg-b, 14-68)

77-C. (Russian) INFLUENCE OF GAS PHASE COMPOSITION ON REDUCTION KINETICS OF FLUXED AGGLOMERATE SINTER. P.N. Ostrik and S.T. Rostovtsev. Izvestiya VUZ--Chernaya Metallurgiya, no 9, 1962, p 17-25.

Reduction rate of sintered iron ore cake with given percentages of Fe, FeO, CaO, SiO_2 and CaO/SiO_2, at stable 800° C. temperature, as influenced by sinter porosity, basicity and amount of H_2, CO and N_2 added to natural gas fuel. 8 ref. (C26, 2-60, 2-61; Fe-b)

78-C. (Russian) REDUCTION OF MOLTEN COPPER-OXIDE BY SOLID CARBON. V. V. Kondakov, D.I. Ryzhonkov and I.A. Titova. Izvestiya VUZ--Chernaya Metallurgiya, no 9, 1962, p 26-30.

Determination of the reduction rate, dissociating energy and activation energy of Cu_2O tablets heated at 1350 to 1520° C. in contact with graphite crucible and ZrO_2 crucible. 4 ref. (C26, 2-62; Cu-b, 14-68)

Fig. 6.1. *(Continued)*

35 0508 **Study of [the Corrosion-Resistance and Properties of] Two Aluminium Alloys Used by Japanese Builders of Liquefied-Natural-Gas Tanks and Tankers.** Walter Trüb. Schweiz. Aluminium Rundschau, Feb. 1969, 19, (1), 28-34 [in German and French].
Laboratory and $\frac{1}{5}$th- and full-scale tests at +20 and —196 °C on Peraluman-460- and Unidur-100-type (Al—4.5% Mg—Mn and Al—4% Zn—Mg) alloys used in Japan for construction of tanks and tankers for storage and sea transport of liquefied gas are reported. Resistances of the structures to sea-water and stress corrosion, and to fatigue of welded and unwelded plates and sections, are compared, and UTS, YP, and extension data are plotted and tabulated.—J. R.

35 0509 **[Comment on a Paper by W. Friehe, B. Meuthen, and W. Schwenk on:] Weathering Behaviour of Aluminized and Galvanized Steel Sheet in Relation to Aluminium and Zinc in Industrial, Rural, and Marine Atmospheres.** Konrad
- Primke. Werner Friehe, Bernd Meuthen, and Wilhelm Schwenk. Stahl u. Eisen, 23 Jan. 1969, 89, (2), 88-89 [in German]. (x).
Two letters. The original investigation (ibid., 1968, 88, 477; Met. A., 6809-35 0881) which was based on an identical coating thickness of 45 μm normally attainable with hot galvanizing and aluminizing is attacked by P. as being biased against metal spraying which requires a min. coating thickness of 80 (preferably 100 μm) for Zn and 200 μm for Al. The benefits of spray-deposited metal coatings are listed: (i) the equivalent of hot-dip galvanized and roll-clad coatings; (ii) independence of the size of part requiring protection; and (iii) the wide choice of spraying materials (including ceramics); examples of the durability of metal-sprayed deposits are quoted. F., M., and S. in reply stress the object of the above original investigation which was to define the limits of the different protection methods on exposure to various aggressive media, for which one of the parameters—in this case coating thickness—was initially kept constant. The results, they conclude, do not detract from the characteristic advantages of metal-sprayed deposits. 17 ref.—H. E. W.

35 0510 **Corrosion of Refractory Ceramic Materials in the Plasma of Combustion Gases.** P. Hoch and M. Masek. Werkstoffe u. Korrosion, Dec. 1968, 19, (12), 1026-1031 [in German].
The corrosion resistance of phosphate cements and of refractory ceramics based on Al_2O_3, MgO, and ZrO_2 in flowing ionized gases (air, CO_2, H_2O, and N_2) was investigated at temp. up to 1800 °C. Sintered ZrO_2 and spinel do not withstand water vapour and impact satisfactorily under these conditions. The specimens were weighed before and after each test, which lasted for 5 min. An experimental qualitative interpretation of the phenomena was obtained on a basis of available thermodynamic and chemical data. In the case of Al_2O_3 binders, a high P_2O_5 content in the mass can reduce resistance to hot gas erosion. 21 ref.—H. S.

35 0511 **Corrosion of Brass in Ammonium Chloride Solution.** R. Bartonicek, M. Holinka, and M. Lukasovska. Werkstoffe u. Korrosion, Dec. 1968, 19, (12), 1032-1042 [in German].
The corrosion performance of brass (77% Cu—21% Zn—2% Al with and without addition of As) was investigated in 0.01-1M ammonium chloride solution of pH 4 and 2, and a solution contg. 1 mole of Cl⁻ with variable Na⁺ and NH_4^+ concentrations at pH 4 and 2. Intercrystalline corrosion was found in brasses contg. as little as 0.01% As. As-free

Fig. 6.2. Examples of informative abstracts. *Source. Metals Abstracts,* **1,** June 1969. Reprinted with permission.
Note: The reader will observe that some abstracts are more fully informative than others.

brasses were attacked by dezincification. In the presence of M-Cl^-, intercrystalline corrosion occurred at pH 2 irrespective of the ammonium ion concentration, but only at much higher ammonium ion concentrations at pH 4. The corrosion was caused by Zn and As segregation at the grain boundaries, and to the influence of As on the stability of the $CuCl_2^-$ formed by the anodic reaction. 49 ref.—H.S.

35 0512 **Protection of Boilers from Corrosion.** M. Postelnicescu and C. Neaga. Bul. Inst. Politehn. Bucuresti, May-Aug. 1968, **30**, (3/4), 187-205 [in Rumanian].
The protection of boilers and heat-exchangers operating under sulphurizing conditions is examined, and the possibilities of reducing the fuel consumption to the min. are discussed.—L.R.

35 0513 **Problems of Cavitation [in Relation to the Erosion of Steels and Other Metals].** S. Nemechek. Elektronnaya Obrabot. Mat., 1968, (6), 50-56 [in Russian].
A number of conference papers relating to the cavitation in liquids and the effect of this phenomenon on the mechanical properties and erosion resistance of steels and other metals used in hydraulic equipment are reviewed and discussed. Special attention is paid to the measures which may be taken in order to reduce cavitation erosion in turbines and associated equipment. The physical principles underlying the development of cavitation erosion are explained and the part played by electrochemical corrosion in the failure of metal parts used under such conditions is indicated.—G.A.

35 0514 **Contact Interaction Between Silicon Carbide and Molten Copper.** G.G. Gnesin and Yu. V. Naidich. Poroshkovaya Met., Feb. 1969, (2), 57-63 [in Russian].
Contact interactions taking place between molten Cu and the surface of SiC single crystals and polycrystalline aggregates were studied, with special ref. to the wetting characteristics of the metal. Pure Cu decomposed the SiC and formed a Cu—Si alloy together with free graphite; this reaction took place far less vigorously in the case of Cu samples contg. traces of Si. Similarly, SiC contg. traces of free Si was more resistant to the action of Cu than the pure carbide. The mechanisms responsible for these effects are discussed. 11 ref.
—G.A.

35 0515 **Corrosion of Boron-Metal Composites.** M.C. Porter and E.G. Wolff. Paper from 'Advances in Structural Composites'. V. 12. 1967, 16 p (AC-14) (Met. A., 6906-72 0082) [in English].
Four systems, Al—B, Al alloy 40E/B, Cu—B and Ni—B, were tested in corrosive environments including solutions of nickel chloride, sodium chloride, ferric chloride and aluminum sulfate. The rates of corrosion under varying conditions of temperature and aeration were measured for these systems. In addition, the effect of corrosion on the mechanical properties of Al—B composites was evaluated. In general, the presence of B resulted in only slight increases in the corrosion rates measured.—AA

35 0516 **The Oxidation of Metals Containing Dispersed Oxide Particles.** R.A. Rapp. Paper from 'Oxide Dispersion Strengthening'. 1968, 539-562 (Met. A., 6906-72 0083) [in English].
Reviews the isothermal oxidation kinetics, scale morphologies and scale adherence for the oxidation of metals containing dispersed oxide particles. Both small, equiaxed dispersed oxide particles (inert and

Fig. 6.2. *(Continued)*

reactive) and larger lamellar particles are considered in combination with matrix metals which form scale principally at the oxide/gas interface, as well as metals which form scale at the metal-scale interface. Selected pertinent investigations from binary alloy oxidation have been chosen for discussion to complement the rather limited observations available for the oxidation of dispersion strengthened metals. Although the oxidation behavior of TD Ni and similar dispersion strengthened metals differs little from that of the matrix metal, other types of metal/particle combinations can result in significantly different oxidation behavior. Further experiments are needed to substantiate the proposed models and their influence on the oxidation behavior. 41 ref.—AA

35 0517 **New Automatic System Optimizes Cooling Tower Corrosion Control.** H. Feitler and C. R. Townsend. Mater. Protect, Mar. 1969, **8**, (3), 19-22 [in English].
To economize on costs of heat removal in cooling systems, cooling water often is re-used; the heat pickup is frequently removed by evaporation. As water evaporates, the total dissolved solids (TDS) of the remaining water increase, causing scaling and corrosion. For max. re-use cycles of the water, its TDS and pH must be kept in proper balance. However, scale can form on hot spots and, for this reason, scale inhibitors are used. An automatic control system for optimizing corrosion and scale prevention in cooling towers maintains, within preset limits, the water's pH, TDS, inhibitor level and corrosion rate. The operation of the system, its advantages, installation, maintenance and economics are discussed.—M. J. C.

35 0518 **Corrosion of Metals in Tropical Environments—Nickel and Nickel—Copper Alloys.** C. R. Southwell and A. L. Alexander. Mater. Protect, Mar. 1969, **8**, (3), 39-44 [in English].
Corrosion data are reported for Ni and Ni—Cu alloys exposed for 16 years to fresh water, sea water, sea water mean tide, and marine and inland atmospheres in the Panama Canal Zone. Major emphasis is on the sea water environments. Results reported include weight loss, pitting and change in tensile strength for simple plates and weight loss for galvanic couples. Marine corrosion in the tropics is compared with available Ni alloy results from temperature latitudes in the U.S.. Time vs weight loss and time vs pitting curves are presented. A few other nonferrous metals are included for comparison. 13 ref.—AA

35 0519 **The Corrosion of Zinc in KOH Solutions.** T. P. Dirkse and R. Timmer. J. Electrochem. Soc., Feb. 1969, **116**, (2), 162-165 [in English].
The corrosion rate of Zn in KOH solutions has been measured under a variety of conditions. Amalgamation and the presence of zincate ions lower this rate of corrosion. The effect of increasing KOH concentration on the rate of corrosion is different for nonamalgamated Zn than for amalgamated Zn. The temperature effect is also different for the two types of Zn electrodes.—AA

35 0520 **Mechanism of Inhibiting Stress Corrosion Cracking of 18-8 Stainless Steel in MgCl₂ by Acetates and Nitrates.** H. H.

Fig. 6.2. *(Continued)*

Uhlig and E. W. Cook, Jr. J. Electrochem. Soc., Feb. 1969, 116, (2), 173-177 [in English].
Modest additions of sodium acetate, nitrate, iodide, or benzoate to $MgCl_2$ test solution, boiling at 130 C., are found to increase resistance to or inhibit stress corrosion cracking of 18-8 stainless steel. The critical applied potential in $MgCl_2$ solution (-0.145 V) above which, but not below, cracking occurs, is shifted in the noble direction by extraneous salt additions. When the shift exceeds the corrosion potential for 18-8 in the same solution, cracking is apparently inhibited. On the other hand, salt additions, $(FeCl_3)$ which shift the corrosion potential in the noble direction may induce or accelerate stress corrosion cracking. The critical potential is interpreted as that value, above which but not below, Cl^- ions adsorb to cause imperfection sites of plastically deforming metal in an amount adequate to cause failure (stress sorption cracking). The present data do not support an electrochemical mechanism of stress corrosion cracking based on anodic dissolution of metal ions at the tip of a crack, nor the mechanism dependent on continuous cracking of a surface oxide film. 18 ref.—AA

35 0521 An Electrochemical Mass Transport—Kinetic Model for Stress Corrosion Cracking of Titanium. T. R. Beck and E. A. Grens, II. J. Electrochem. Soc., Feb. 1969, 116, (2), 177-184 [in English].
A quantitative model for the electrochemical kinetic and mass transport process in a propagating stress corrosion crack was used to gain insight into the manner in which these processes influence propagation. Analysis of the problem led to system of simultaneous differential equations which, with their appropriate boundary conditions, were solved by computer implemented numerical methods. Comparison of computed behavior with experimental stress corrosion cracking data for a Ti alloy has guided the development of the model and the specification of critical stress corrosion cracking experiments. Such comparisons indicate that there is a halide ion current to the crack tip with some H_2 ion discharge in the region downstream from the tip. A significant fraction of the current entering a crack appears to be involved in the formation of soluble Ti ions parallel with oxide formation on the walls. 11 ref.—AA

35 0522 Methods for Studying the Solution Chemistry Within Stress Corrosion Cracks. B. F. Brown, C. T. Fujii, and E. P. Dahlberg. J. Electrochem. Soc., Feb. 1969, 116, (2), 218-219 [in English].
Specimens of 7075 Al alloy, Ti—8%Al—1%Mo—1%V alloy and vacuum melted C-deoxidized 0.45% C steel were notched, stressed and placed in water or $3\frac{1}{2}$% NaCl solution until stress corrosion cracks had propagated for a suitable distance. Specimens were then placed in liquid N_2 to freeze the corrodent in the crack. They were broken apart and analyzed immediately on thawing with indicator-impregnated filter paper or indicator-coated silica gel to determine the acidity and soluble metallic constituents. The region of highest acidity was found at the advancing crack edge with a gradual decrease toward the crack root. Results suggest that the acidity is the result of hydrolysis of the metallic constituent relased by anodic dissolution.
—M. J. R.

35 0523 High-Temperature Oxidation of Co-10 w/o Cr Alloys. Pt. 1. Microstructure of Oxide Scales. P. K. Kofstad and A. Z. Hed. J. Electrochem. Soc., Feb. 1969, 116, (2), 224-229 [in English].
Oxidation of Co-10 wt% Cr alloy results in an outer CoO layer, an inner CoO layer with inclusions of Cr_2O_3 and $CoCr_2O_4$, and an inter-

Fig. 6.2. *(Continued)*

nal oxidation zone. Both oxide layers contain appreciable porosity and the morphology of the inclusions is a function of temperature. The microstructure of scales formed during oxidation is presented. It is assumed that the rate-controlling step is the solid state diffusion of Co ions in the CoO phase, while gaseous O_2 is rapidly transported across pores. The pores short-circuit the solid state diffusion process and enhance oxidation, while Cr_2O_3 and Co—Cr spinel inclusions inhibit the oxidation. 17 ref.—AA

35 0524 **High Temperature Oxidation of Co-10 w/o Cr Alloys. Pt. 2. Oxidation Kinetics.** P. K. Kofstad and A. Z. Hed. J. Electrochem. Soc., Feb. 1969, 116, (2), 229-234 [in English].
Thermogravimetric studies of the oxidation kinetics of Co-10 wt% Cr have been carried out in the temperature range 800 to 1300 C. at O_2 pressures from 0.05-760 torr. The over-all oxidation is parabolic, and the O_2 pressure dependence of the rate constant can be expressed by $k_p \propto p_{O_2}^{1/n}$, where n varies from 3 at 900 to 2.5 at 1300 C. The oxidation is faster than that of pure Co. The oxide scale is double-layered and consists of an outer CoO layer and an inner layer of $CoCr_2O_4$ and Cr_2O_3 particles embedded in a CoO matrix. The inner layer also contains 30-35 vol % porosity. It is concluded that the oxidation is controlled by Co-vacancy diffusion in the CoO-phase. The spinel inclusions inhibit the oxidation by decreasing the effective diffusion area in the scale. The pores partially short-circuit the solid-state diffusion through the scale due to O_2 transport across the pores; the porosity then serves to increase the oxidation. Semiconductor valence effects due to dissolution of Cr in the CoO phase are concluded to be of minor importance with regard to the relative oxidation behavior of pure Co and Co-10 wt% Cr alloy. 23 ref.—AA

35 0525 **Study on Electrolytic Machining Process. XII.—Corrosion of Mild Steel in Sodium Chloride Solution (II).** Sukemitsu Ito, Hideo Yamamoto, and Kunio Chikamori. J. Mech. Lab. Japan, 1967, 13, (1), 15-24 [in English].
See J. Mech. Lab. (Tokyo), 1967, 21, (3), 103; Met. A., 6805-35 0426. See also Met. A., 6902-53 0025.

35 0526 **The Effect of Weathering on Hot-Galvanized High-Tension Pylons after 20-30 Years Exposure.** J. F. H. van Eijnsbergen. Jahrbuch Oberflächentechnik 1969, 1969, (25), 223-226 (Met. A., 6903-53 0049) [in German]. (x).
A description is given of the results of observations made on hot-dip galvanized electricity pylons after some 18 years exposure to the windy seaside atmosphere at Den Helder, the industrial atmosphere at Ymuiden, and the general polder environment of North Holland as a whole. The examinations show the side opposite the wind yields specimens which have a pure Zn layer still partly unaffected, and a fully intact steel base metal. The specimens taken from the side exposed to the wind however showed extremely severe corrosion due to the weathering effect of several years exposure. Similar observations made on pylons in rural Friesland show some to be intact after 30-35 years whereas in the more industrialized South Holland, the atmosphere had caused severe corrosion after only 12 years. The mechanism of rust patch formation in the connecting plates is described and illustrated in cross-section.—P. R. L.

35 0527 **The Formation of Diffusion Layers in the Oxidation of Some Transition Metals.** L. F. Sokiryansky, V. V. Latsh, S. S. Mozhaev, L. G. Maksimova, and N. F. Kharlamova.

Fig. 6.2. *(Continued)*

'Diffuziya v Metallakh i Splavakh' (Russian Collection, Tul'sk. Politekhn. Inst., Tula) 1968, 367-374 [in Russian, R-Z]. The interpretation of results of a kinetic study of the process of oxidation in the Ti—O system is discussed on the basis of a proposal that there is formation of the ordered compounds Ti_6O and Ti_3O. The O concentration was calculated from the c value of the Ti lattice. For all cases of gas saturation, X-ray data underestimates the true value of the O concentration in the surface being analysed. A calculation was made of the reflection profiles as a function of the concentration gradient. The form of the reflections for two values of equilibrium concentrations of O (17 and 29 at.-%) were compared with the nature of the O distribution in the surface layer with various values of the diffusion coeff. Analysis of experimental data showed that X-ray analysis gives more complete information about the structure of gas-saturated layers than do the microstructure, microhardness, and other methods. It is stated that there is no conclusive proof that an ordering process occurs in the diffusion zone.—A. D. M.

35 0528 **The Corrosion and Mechanical Properties of Zirconium— Beryllium—Tin Alloys.** A. S. Adamova and A. T. Grigor'ev. 'Fiziko-khimiya Splavov Tsirkoniya' (Russian Collection, Published 'Nauka', Moscow) 1968, 64-70 [in Russian, R-Z]. See also ibid., p 71; Met. A., 6906-31 0987. Results are given for the corrosion and mechanical properties of alloys of the Zr corner of the Zr—Be—Sn system for four radial sections with Sn : Be = 3 : 1, 1 : 1, 1 : 3, and 1 : 9, resp., for (Sn + Be) up to 11-14%. Cast alloys of (Sn + Be) ≤ 2-4% had good corrosion resistance in water at 350 °C and 168 atm pressure, i.e, 0.0002-0.0007 g/m² h. Low-alloy forged alloys also had satisfactory corrosion resistance in these conditions. The rate of oxidation of all the ternary alloys in air at 650 °C was greater than that for unalloyed Zr. The strongest alloys at room temp. and 400 °C were found. The alloy compositions with good corrosion and mechanical properties were: Zr—1.50% Sn—0.50% Be, Zr—1.50% Sn—1.50% Be, Zr—0.25% Sn—0.75% Be, Zr—0.50% Sn—1.50% Be, and Zr—0.20% Sn—1.80% Be.—A. D. M.

35 0529 **The Corrosion and Mechanical Properties of Zirconium— Iron—Nickel Alloys.** E. M. Tararaeva and A. T. Grigor'ev. 'Fiziko-khimiya Splavov Tsirkoniya' (Russian Collection, Published 'Nauka', Moscow) 1968, 113-117 [in Russian, R-Z]. The corrosion and mechanical properties of Zr-based alloys contg. (Ni + Fe) 0.5-2.0% in the ratios of 1 : 2, 1 : 1, and 2 : 1 were studied. Corrosion tests with Zr—Fe—Ni alloys in a water—steam mixture at 350 °C and 170 atm pressure showed that their corrosion resistance was satisfactory over 3500 h although it fell after a further 500 h. Corrosion tests in CO_2 gas at 500 °C and 20 atm pressure for 2000 h showed that the combination of Fe + Ni did not improve the corrosion resistance of Zr. Oxidation studies in air at 650 °C for 400 h showed that Fe + Ni improved the resistance to oxidation, increasing additions improving the resistance. Tensile tests at room temp. and 400 °C showed that the addition of (Fe + Ni) 1.5% to Zr raised the UTS by 1.5 times. Fe + Ni also improved the creep resistance of Zr at 400 °C significantly.—A. D. M.

35 0530 **The Corrosion Resistance of Zirconium—Copper—Nickel Alloys in Various Media at Elevated Temperatures.** N. V. V'yal' and O. S. Ivanov. 'Fiziko-khimiya Splavov Tsirkoniya' (Russian Collection, Published 'Nauka', Moscow) 1968, 164-169 [in Russian, R-Z]. Corrosion tests of Zr—Cu—Ni alloys were made in (i) water, (ii) 3% H_2SO_4 + 3% UO_2SO_4, and (iii) CO_2 gas. Alloys of Zr contg. (Cu + Ni) 0.25-3.0 at.-% have satisfactory corrosion resistance to (i) at 350 °C

Fig. 6.2. *(Continued)*

58

(5000 h) that is better than that of Zr—Mo—Cu or Zr—Fe—Ni alloys. The most resistant alloy was Zr—0.42 wt.-%Cu—0.10 wt.-%Ni which had a weight gain after 5000 h of 2.6 g/cm^2. Zr with low alloy contents of (Cu + Ni) 0.2-1.0 wt.-% was corrosion resistant in (ii) at 300 °C and 87 atm pressure in a 2000 h test. The Zr—Cu—Ni alloys were more resistant to (iii) than Zr—Mo—Ni or Zr—Fe—Ni alloys with the Ni having more effect than Cu on the corrosion resistance. However, the Zr—Cu—Ni alloys after a 2000 h test had an increase in weight although showing no signs of external attack.—A. D. M.

35 0531 **The Corrosion Properties of Zirconium—Molybdenum—Niobium Alloys.** N. M. Gruzdeva and A. S. Adamova. 'Fiziko-khimiya Splavov Tsirkoniya' (Russian Collection, Published 'Nauka', Moscow) **1968**, 204-208 [in Russian, R-Z].
A study was made of the corrosion resistance of ternary Zr—Mo—Nb alloys in air at 650 °C and in water at 350 °C and 168 atm pressure. Low alloys (<1.0% Mo + Nb) of three radial sections with Nb:Mo ratios of 4:1, 1:1, and 1:4 were studied after quenching from 1200 and 600 °C. The heat resistance of all the ternary alloys in air at 650 °C is lower than that of unalloyed Zr. The Zr—0.80%Nb—0.22% Mo and Zr—0.50%Nb—0.56%Mo alloys quenched from 1200 °C had good corrosion resistance in water at 350 °C and 168 atm pressure, i.e. ~0.004 g/m^2 h after 2225 h. The low-alloy Zr alloys quenched from 600 °C do not have a good resistance to corrosion in water at 350 °C and 168 atm pressure.—A. D. M.

35 0532 **The Effect of Silicon, Tin, and Chromium on the Corrosion and Mechanical Properties of Zirconium—Molybdenum—Niobium Alloys.** N. M. Gruzdeva and A. S. Adamova. 'Fiziko-khimiya Splavov Tsirkoniya' (Russian Collection, Published, 'Nauka', Moscow) **1968**, 208-215 [in Russian, R-Z].
A study was made of the effects of Sn 0.2-0.3, Cr 0.1, and Si 0.1% on the corrosion and mechanical properties of the ternary alloys Zr—0.80% Nb—0.20% Mo; Zr—0.50% Nb—0.20% Mo; and Zr—0.50% Nb—0.50% Mo. This type of alloying decreases the heat resistance of the ternary alloys at 650 °C in air. Addition of Cr 0.1% to Zr—0.50% Nb—0.2-0.5% Mo improves the corrosion resistance of cast or forged alloys to water at 350 °C and 168 atm pressure and to steam—water at 400 °C and 250 atm pressure. Sn and Si lower the corrosion resistance of all the alloys. The alloys with the highest strength and creep resistance are those based on the ternary Zr—0.50% Nb—0.20-0.50% Mo alloyed with Cr 0.1 and Sn 0.2%.—A. D. M.

35 0533 **The Effect of Iron, Nickel, and Chromium on the Corrosion and Mechanical Properties of Zirconium—Molybdenum—Niobium and Zirconium—Copper—Tin Alloys.** N. M. Gruzdeva and A. S. Adamova. 'Fiziko-khimiya Splavov Tsirkoniya' (Russian Collection, Published 'Nauka', Moscow) **1968**, 215-222 [in Russian, R-Z].
A study was made of the effect of (i) Cr 0.2-0.3, Fe 0.2-0.3, and Ni 0.1-0.2% on the corrosion and mechanical properties of forged Zr—0.80% Nb—0.20% Mo and Zr—0.50% Nb—0.20% Mo alloys and (ii) Cr 0.1, Fe 0.2, and Ni 0.1% on the same properties of forged Zr—0.67% Sn—1.33% Cu and Zr—1.50% Sn—1.50% Cu alloys. Alloying of Zr—0.50% Nb—0.20% Mo with Cr 0.2-0.3% and Fe 0.2-0.3% improved its corrosion resistance in water at 350 °C giving corrosion rates of 0.001-0.002 g/m^{2}h after 4000 h. The corrosion resistance of the ternary Zr—Cu—Sn alloys was improved by addition of Fe 0.2% but lowered by Cr and Ni. Additions of Cr and Fe strengthened the ternary alloys.—A. D. M.

Fig. 6.2. *(Continued)*

35 0534 The Corrosion Resistance of Alloys of the Zirconium–Niobium–Chromium System. N. M. Gruzdeva, T. N. Zagorskaya, and I. I. Raevsky. 'Fiziko-khimiya Splavov Tsirkoniya' (Russian Collection, Published 'Nauka', Moscow) **1968**, 251-257 [in Russian, R-Z].

Data are given for the corrosion resistance of Zr–Nb–Cr alloys in (i) water at 350 °C and 169 atm pressure and (ii) air at 650 °C for 20 h. The alloys tested were those of the Zr corner of the system (Nb + Cr) up to 50%. After a 2184 h test in (i) the alloys showing the highest corrosion resistance were those of the Nb : Cr = 3 : 1 section contg. (Nb + Cr) 1-2% and 15-17%, the weight increments being 7. 5-9. 5 and 2. 5-3. 0 g/m² resp. Alloys contg. (Nb + Cr) 40 and 50% disposed along the same section also had good corrosion resistance (3-5 g/m²). Tests in (ii) showed all the alloys to have low resistance except for the Zr contg. low-alloy additions of (Nb + Cr) 1. 0% and 2. 0% from the sections Nb : Cr = 1 : 3 and 1 : 1, resp. The oxidation resistance of these alloys in a 20 h test was higher than that for unalloyed Zr.—A. D. M.

35 0535 The Corrosion Resistance of Zirconium Alloys in Super-heated Steam at 500 °C and 100 Atmospheres Pressure. N. V. V'Yal' and O. S. Ivanov. 'Fiziki-khimiya Splavov Tsirkoniya' (Russian Collection, Published 'Nauk:', Moscow) **1968**, 257-261 [in Russian, R-Z].

Results are given for the resistance of Zr alloys to corrosion in superheated steam at 500 °C and 100 atm pressure. Binary alloys of Zr contg. Nb, V, or Ta 0. 5-1. 5 at. -% and ternary and quaternary alloys with these elements present as a sum total of 1. 0 at. -% did not give satisfactory corrosion resistance, i.e. weight increases of 81-196 g m² in 400 h tests. Ternary alloys of Zr contg. Cu, Cr, or Fe to a sum total of 2. 0 at. -% showed increased corrosion resistance, e.g. after 4000 h the weight gains of Zr–Cu–Cr and Zr–Cu–Fe alloys were 47-51 and 29-37 g/m², resp. Ternary alloys of Zr with (Nb + Ta) 25 and 50 at. -% failed after 250-1000 h testing. A study was also made of the effect of Pd on the corrosion resistance of some Zr alloys.
—A. D. M.

35 0536 Interaction Between Metals and the Hydrides of Non-Metals. T. M. Mikhlina, V. A. Obolonchik, and M. M. Antonova. Ukrain. Khim. Zhur., Feb. 1969, **35**, (2), 128-130 [in Russian].

The effect of the hydrides of non-metals, for example, H₂Se, on rare-earth metals and their hydrides was studied. At the onset of the inter-action between La, Re, and similar metals and H₂Se, H was evolved from the latter as a result of thermal dissociation, and this had a substantial effect on the physical and chemical properties of the metals. It was very difficult to obtain metallic compounds active with respect to H by this method, as the initial temp. corresponding to the interaction exceeded the temp. corresponding to the formation of the metal hydrides. 9 ref.—G. A.

35 0537 Effects of Composition on the Stress Corrosion Cracking of Ferritic Stainless Steels. A. P. Bond and H. J. Dundas. Climax Molybdenum Co., 1270 Ave. of the Americas, New York, N. Y. 10020. (Reprint from Corrosion, Oct. 1968, **24**, (10), 334-352). 1968, 10 p. [Pamphlet—English].

U-bend and axially loaded tensile-type specimens were used in stress corrosion tests performed in boiling 140 C. (284 F.) magnesium chloride solution. The experimental ferritic alloys tested contained 17-25% Cr, 0-4% Ni, 0-2% Cu and 0-5% Mo. Alloys essentially free

Fig. 6.2. *(Continued)*

of Ni and Cu did not undergo stress corrosion cracking (tests on 430 and 434 stainless steels confirmed this behavior). However, alloys containing more than 1.0% Ni or 0.5% Cu were subject to transglandular stress corrosion cracking in boiling magnesium chloride solution. Critical concentrations of Cu and Ni that produce susceptibility to cracking were determined at various Cr and Mo levels. Experiments at controlled potentials indicated that cracking in boiling magnesium chloride was not the result of H_2 embrittlement. 17 ref.—AA

35 0538 **Here's How to Tackle Corrosion by Acids.** J.D.Palmer. Can.Chem. Process., Jan.1969, **53**, (1), 50-53 [in English].
Metallic materials offering corrosion resistance to a wide variety of acids, nonmetallic materials for use in conjunction with metals for additional protection, and chemical and thermal means of combatting corrosive attack by acids in special situations are reviewed. The curves and diagrams used to predict corrosion rates under specific conditions of temperature, current, fluid velocity, aeration and trace impurities are illustrated and discussed. For example, the corrosion rates of Al in solutions of lactic, succinic, malic, and gluconic acids, as a function of acid concentration, are shown in a set of curves.
—S.M.

35 0539 **The Oxidation of Tantalum Metal in Carbon Dioxide Atmospheres.** K.J.Richards and M.E.Wadsworth. [Met.Soc. Conf.] High-Temperature Refractory Metals, New York, 1965, 1969, **34**, (1), 291-304 (Met.A., 6904-72 0061) [in English].
The oxidation of Ta by CO_2 at various pressures was studied in the range 700-950 °C, using a thermobalance. Oxidation in the temp.range 720-830 °C is linear and is controlled by an interfacial reaction associated with the formation of a non-protective Ta_2O_5 layer, with an activation energy of 40.4 kcal/mole. Above and below this temp. range the reaction mechanism becomes more complex and analysis is difficult. Variation in the colour of the oxide with the temperature of its formation is discussed in terms of departure from stoichiometry.—J.F.T.

35 0540 **Oxidation Kinetics of Tantalum Carbide.** Ronald L.Gibby and Milton E.Wadsworth. [Met.Soc.Conf.] High-Temperature Refractory Metals, New York, 1965, 1969, **34**, (1), 305-324 (Met.A., 6904-72 0061) [in English].
Powdered TaC was oxidized under different O partial pressures at 600-825 °C. Linear oxidation occurs in the temp. range 650-825 °C, the mechanism involves two parallel linear rate-controlling reactions, the predominance of either depending upon the temp. The activation enthalpies involved are 108 and 55 k cal/mole, and the oxide formed is a relatively non-protective Ta_2O_5 layer. Below 650 °C parabolic oxidation is observed, which involves the formation of an intermediate compound designated $Ta(C_xO_y)$ with an activation enthalpy of 55 kcal/mole. 9 ref.—J.F.T.

35 0541 **Operating Experience and Corrosion Preventive Measure of Residual Fuel Firing Gas Turbine.** Shichinosuke Tanaka et al. Mitsubishi Tech.Rev., Sept.1968, **5**, (3), 196-207 [in English].
A residual oil contg. V 25-45, Na 30-50 ppm and S 3% was fed to a gas turbine. During the $1\frac{1}{2}$ year operation the unit was overhauled every 3 months and no indication of corrosion on the hot section blades was found. The state of corrosion, the deposits on the turbine blades, and the process of change in properties of the blade materials in corrosion were also discussed. 10 ref.—L.R.

Fig. 6.2. *(Continued)*

61

35 0542 On the Structure of Oxide Films Found on Tantalum Surface by Wet Oxidation. Tadayuki Nakayama and Toshiaki Osaka. Rep. Castings Research Lab. Waseda Univ., Nov. 1967, (18), 29-33 [in English].

99.97% Ta samples were heated in an autoclave in water or water vapour to 300 °C for 1 or 24 h. The resultant surface films were studied by transmission electron diffraction. Both films were of the rutile type; that formed in water was identified as topiolite, $(Fe, Ta)O_2$ and that in water vapour as TaO_2. 13 ref.—P.C.K.

35 0543 Prozamet's Contribution to Development of Anti-Corrosive Coatings. Miroslaw Wielicki. Problemy Projektowe Hutn., June 1968, **16**, (6), 196-200 [in Polish].

One third of world's annual steel production is lost through corrosion. The significance of protective measures is discussed and the activities of Prozamet in this field since 1951 are surveyed. Its principal contributions were studies of paint and metallic coatings and methods of application. This involved field studies and devising and developing protective schemes and advisory services as well as design and construction of apparatus, plant, and factory equipment, including planning, installation and/or modernisation of over 150 paint and galvanizing works.—B.J.S.

35 0544 The Corrosion Resistance of Arsenical Steel. A.V. Tursunov and V.L. Gutorova. Sbornik Trudy Donets. Nauchn.-issled. Chern. Met., **1968**, (5), 141-144 [in R-Z].

Studies were made of the corrosion resistance of a steel contg. C 0.17-0.21, Si 0.14, Mn 0.44-0.49, S 0.029-0.048, P 0.022-0.042, and As 0.115-0.156%. The presence of 0.16% As in low-C steel has no deleterious effect on its corrosion resistance in mains water (it may even improve it to some extent) and does not lower the resistance of the steel to gaseous corrosion at 350-700 °C. In terms of corrosion resistance, As steel can be used as well as C steel in boiler construction.
—A.D.M.

35 0545 The Heat [Oxidation] Resistance of Nickel-Based Alloys. A.V. Guts, L.A. Panyushin, and A.S. Tumarev. Trudy Leningrad Politekhn. Inst., 1968, **(295)**, 90-97 [in Russian, R-Z].

A review is given of theoretical ideas on the nature of gaseous corrosion of alloys at high temp. Results are given from a study of the oxidation kinetics of Ni—Cr alloys. The diffusion processes in the scale are the limiting stage in the oxidation process of these alloys and the heat resistance of these alloys varies over a wide range depending on their composition and the structure of the scales formed. A study was also made of the oxidation of Ni—Al, Ni—Ti, and Ni—Co alloys. All alloys showed an overall parabolic oxidation law with time indicating the determining role of the diffusion of reagents in the scale. 16 ref.
—A.D.M.

35 0546 The Cathodic Reduction of Some Inhibitors of the Corrosion of Titanium. A.P. Brynza, V.P. Fedash, V.I Sotnikova, Yu.M. Loshkarev, and T.P Legashova. Zashchita Metallov, 1969, **5**, (1), 10-14 [in Russian].

Cathodic polarization curves taken at a rate of 0.1-1.0 V/min from an initial potential of +0.4 V were obtained for Ti (VT1-1, ⪕ 99.25% Ti) specimens in $5N$-H_2SO_4 contg. various organic corrosion inhibitors, i.e. (i) p-nitroaniline, (ii) p-nitrobenzaldehyde, (iii) α-nitroso-β-naphthol, (iv) nitrobenzene, (v) neutral red, and (vi) methyl orange, all of which shift the steady state potential of Ti in H_2SO_4 to values of +0.1 to +0.4 V. Polarization curves with a clearly defined current plateau were obtained in all cases indicating that inhibitor reduction

Fig. 6.2. *(Continued)*

was diffusion controlled, e.g. the current values increased with increase in inhibitor concentration and with stirring; activation energies of 4.6-5.0 kcal/mole, typical for diffusion, were obtained. The inhibitors (v) and (vi) were effective only on partially oxidized surfaces and did not protect an active Ti surface, this result is discussed. The degree of inhibitor efficiency can be determined from the effect of its concentration on the relation of the limiting reduction current to the max. current of active anodic dissolution. 11 ref.—A.D.M.

35 0547 **The Difference in the Corrosion Behaviour of Titanium in Sulphuric Acid and Hydrochloric Acid Solutions.**
A.P. Brynza, L.I. Gerasyutina, E.A. Zhivotovsky, and V.P. Fedash. Zashchita Metallov, 1969, 5, (1), 15-18 [in Russian].
Gravimetric and potentiostatic polarization data were obtained for VT1-1 Ti (<99.25%) to study its corrosion resistance and electrochemical properties in 5-40% H_2SO_4 and HCl solutions in the temp. range 40-80 °C. (Ti is resistant in these acids only at room temp. and in 5-10% concentrations.) At all these temp. the corrosion rate in 5 and 10% H_2SO_4 was greater than in HCl. A graph is given for corrosion rate against the Hammett acidity function. The results show that Cl ions only become more aggressive than SO_4 ions when a certain critical acidity is reached. With increasing temp., H ion activity increases and Ti activation occurs in less acid solutions. A retardation coeff. for Cl ion inhibition of the form $\gamma = V_0/V$ where V_0 and V are corrosion rates in H_2SO_4 and HCl solutions of equal acidity resp. was obtained. The dependence of γ on Cl ion concentration is expressed by a Langmuir type isotherm indicating an adsorption mechanism. γ decreases with temp. and HCl concentration. The adsorption mechanism is discussed in relation to the ϕ scale of potentials. 15 ref.
—A.D.M.

35 0548 **The Protective Action of Nitrogen-Containing and Acetylenic Inhibitors of the Corrosion of Steel in Hydrochloric Acid in Relation to Temperature and Pressure.** N.I. Podobaev and V.V. Vasil'ev. Zashchita Metallov, 1969, 5, (1), 19-26 [in Russian].
A study was made of the most effective N-contg. and acetylenic inhibitors and mixtures of these and of the further addition of I^-, Sn^{2+}, Cr^{3+}, Ni^{2+}, Cd^{2+}, and Al^{3+} ions on the corrosion of St. 1(contg. C 0.07-0.12%) and 36G2S (contg. C 0.32-0.40, Mn 1.5-1.8, and Si 0.4-0.7%) in 4N-HCl in the temp. range 110-250 °C in short-term autoclave tests (this work was in connection with oil-industry requirements for solutions; most acid inhibitors in use have max. effectiveness at 60-90 °C). The inhibitors studied were BA-6 (condensation product of benzylamine and urotropine), PKU (condensation product of urotropine), propargyl alcohol, and hexynol. In general, the Mn steel was more difficult to protect. Increase in pressure decreases inhibitor effectiveness, this effect being studied by capacitance and resistance measurements, and is associated with displacement of adsorbed inhibitor by H atoms. There is some inhibition even at 250 °C although the most effective mixture of BA-6 + Hexynol + KI has the max. effect at 100 °C. Sn^{2+} and Cr^{3+} improve the inhibitive action of acetylenic compounds, and mixtures of these with N compounds, most effectively in the 130-170 °C range. The protective action of inhibitor mixtures in hot acid falls off markedly with time. 15 ref.—A.D.M.

35 0549 **The Corrosion of Aluminium Protected by Polymer Films in Hydrochloric Acid.** Yu. E. Lobanov, A.L. Shterenzon, and L.V. P'yankova. Zashchita Metallov, 1969, 5, (1), 31-38 [in Russian].
A study was made of the corrosion of Al (Al 99.5, Fe 0.26, and Si 0.21%) protected by coatings of polyethylene, an ethylene-propylene

Fig. 6.2. *(Continued)*

copolymer, fluoroplast-26, and fluoroplast-42 (Teflon type) in HCl solutions, by a test method described earlier (Serafimovich and Mikhailovsky, ibid., 1965, 1, 698; M. A., 1966, 1, 583; see also M. *et al.*, ibid., 577; M. A., 1966, 1, 225). The factors studied were polymer adhesion, temp., film thickness, and acid concentration. The steady-state corrosion rate of Al under these films is independent of the adhesion. Over a certain range of acid concentrations, characteristic for each polymer, the corrosion rate is independent of HCl concentration and increases with increasing penetrability of the film—the corrosion process being under mixed diffusion-kinetic control. In more dilute acid the corrosion rate increases exponentially with HCl concentration with the corrosion controlled by HCl diffusion. 9 ref.—
A. D. M.

35 0550 **The Effect of Alternating Current Frequency on the Stray Current Corrosion of the [18 : 9 Chromium—Nickel, Ti-Stabilized] Steel, 1Kh18N9T and of Carbon Steels in Soils and Ground Water.** M. A. Tolstaya. Zashchita Metallov, 1968, 5, (1), 39-46 [in Russian].
Weight-loss and polarization studies were made, in conditions close to those naturally occurring, of the effect of a.c. frequency (0.005-50 Hz) on the stray current corrosion of (i) 1Kh18N9T 18 : 9 Cr—Ni Ti-stabilized steel and (ii) plain-C steel. Tests were made in sandy soils wetted with up to 10-20% of solutions simulating waters of various degrees of mineral content (compositions given). In these conditions the corrosion of (i) decreased sharply with increasing frequency of a symmetrical square-wave polarizing current in the above range. The corrosion of (i) at 0.005 Hz and 100-1000 mA/dm² is more dependent on salt (NaCl) content of the soil than (ii). With low salt contents the corrosion of (i) is only a few per cent of the Faradaic value but increases by ~10 times at higher salt contents. In the presence of industrial currents (50 Hz and 100-200 mA/dm²) steel (i), although having low overall corrosion, is subject to pitting both in aerated and also in very wet soils. The effects of electrode potential amplitude on the corrosion of (i) and (ii) at various frequencies was studied. 29 ref.—A. D. M.

35 0551 **The Stress Corrosion of Heat-Strengthened Reinforcing Steels [For Concrete].** S. N. Alekseev and E. A. Gurevich. Zashchita Metallov, 1969, 5, (1), 47-52 [in Russian].
Anodic polarization studies and stress corrosion tests in $Ca(NO_3)_2$—NH_4NO_3 solution were made with rods of St. 5 contg. C 0.35, Mn 0.68, Si 0.20, S 0.026, and P 0.024% and 35GS contg. C 0.36, Mn 0.98, Si 0.70, S 0.032, and P. 0.012% and a high-strength steel wire contg. C 0.80, Mn 0.91, Si 0.31, Cr 0.13, Ni 0.10, and Cu 0.12%. Tests were made in saturated $Ca(OH)_2$ solution and in concrete with and without additions of 3% CaS or 3% $CaCl_2$. Details of various heat treatments are given. Heat-strengthened reinforcing rods of St. 5 and 35GS under the combined action of stress and specific aggressive agents tend to corrosion cracking and also to failure as a result of the development of cold brittleness. High-strength cold-drawn wire was not susceptible to this type of corrosion under stress. The susceptibility to stress corrosion depends on the metal structure and thus, on the regimes and methods of thermal and mechanical treatment of the reinforcements. 8 ref.—A. D. M.

Fig. 6.2. *(Continued)*

6.4. CRITICAL ABSTRACTS

In a critical abstract the abstractor not only describes the content of the document but also evaluates the work and the way it is presented. This kind of abstract can save the reader time by pinpointing documents of special significance and value.

An example of critical abstracting is found in *Applied Mechanics Reviews* published by the American Society of Mechanical Engineers. This publication provides critical reviews of world literature in applied mechanics and related engineering science. It is the policy of this publication to examine carefully as many primary sources of technical literature as possible and then to select those items that (*a*) have depth and meaning, and (*b*) that appeared in a source that subjects manuscripts to careful refereeing prior to publication. The purpose of the critical review is not to duplicate the refereeing process but to provide an appraisal of the relationship of the selected article to the rest of the literature and to evaluate its worth to the potential reader.

Here are some of the points on which the writer of a critical abstract may wish to comment:

1. Is the document intended for the beginner, the intermediate (or average level) scientist, or the advanced specialist?

2. Is it of wide general interest?

3. Is it oversimplified or overly complex?

4. Are there any prerequisites required for the reader to understand the document?

5. Is the presentation factually accurate?

6. Is the treatment deep or shallow?

' 7. Is it a superficial rehash, a bona fide review article, or are important new points made?

8. Does the author provide sufficient detail?

9. What is the adequacy of the instruments and experimental techniques used?

10. Are there any major typographical errors?

11. Are the experimental results open to interpretations other than that of the author?

12. How does a paper compare with and relate to similar work, and is such work cited?

Some examples of critical abstracts are shown in Fig. 6.3.

At a minimum, the critical abstractor should try to indicate the depth and extent of the work and the level of the audience (elementary, advanced) at which the presentation is aimed. The more advanced and expert the abstractor, the more critical he can afford to be.

There is room for and need for both critical and noncritical abstracting services. But most externally published abstract journals do not permit a critical approach. They may specifically prohibit this in their instructions to abstractors. One reason for this is that the supply of abstractors who have both the ability and the willingness to write critical abstracts is limited. Also, some readers prefer to make their own critiques. Another argument is that many documents have already gone through an extensive editorial review process prior to publication and do not require further criticism. Other reasons are the additional time and funds that are required: good critical abstracts take more time to write and are longer than other kinds of abstracts. Also, a carefully written critical abstract may reach an often impatient audience late, and many fields of science move so rapidly that such delays are not tolerable. Further, each abstractor or reviewer has his own viewpoint, which is not necessarily more valid than that of the author of the original document.

6.5. THE "MINI-ABSTRACT"

An interesting option about which the well-informed abstractor should know is the highly condensed indicative abstract or what we call the *"mini-abstract,"* sometimes also referred to as a "micro-abstract." Client needs may be of such a nature that only a line or so is needed. Also, this approach provides the advantages of speed in both writing and reading, which is often important in fields of high commercial significance and in rapidly moving or highly competitive technical fields. The "mini-abstract" is well suited to operations that have limited funds.

The following are examples of the "mini-abstract":

> Use of plastics in manufacture of XYZ expected to double within the next x years.

> Study of Apollo 11 rock samples shows no evidence of life on the moon.

Another related option is the provision of titles and references (including authors), supplemented by the terms used in indexing. The index terms, especially if the indexing is thorough, can provide additional useful information. For example, see Fig. 6.4. Completeness is sacrificed in the

8701. Szekeres, P., Conformal tensors, *Proceedings of the Royal Society of London, Series (A) Mathematical and Physical Sciences* **304,** 1476, 113–122 (Apr. 1968).

Reviewer cannot help but feel that this paper was written to impress, rather than inform, the reader. The very first equation is written in terms of undefined symbols. The second equation adds to the confusion by introducing undefined and relatively uncommon notation. This situation does not improve throughout paper.

Those readers familiar with tensor notation, on the other hand, will immediately recognize the first equation as the definition of a conformal transformation involving a scalar $\phi(x^i)$, $i = 1, 2, 3$, as given by L. P. Eisenhart [*Riemannian Geometry*, Princeton University Press (1949)], (which is not listed among the references). The notation in the second equation is that occasionally used elsewhere to represent the contravariant derivative, which is related to the covariant derivative. The reader is also expected to know that the covariant derivative of a metric tensor vanishes, and that a comma in the subscript line denotes differentiation with respect to the coordinates whose indices immediately follow the comma. Author employs square brackets to identify the alternation process and round brackets to indicate the mixing process, both explained by Schouten [AMR **8** (1955), Rev. 3593]. This code is not explained in paper.

Author defines a conformal object and a differential concomitant and then shows that a conformal object is a differential concomitant of an object constructed from a metric connection. He also shows that two sequences of conformal tensors, which together generate all conformally-invariant tensors of the space, may be constructed in spaces where the Weyl tensor possesses a nonzero invariant.   W. C. Orthwein, USA

8705. Whiteman, J. R., Treatment of singularities in a harmonic mixed boundary value problem by dual series methods, *Quarterly Journal of Mechanics and Applied Mathematics* **11,** 1, 41–50 (Feb. 1968).

The function u satisfies Laplace's equation in a rectangular region with a slit. The mixed boundary values, consisting of either u or $\dfrac{\partial u}{\partial n}$ given on the boundaries of the region, contain a singularity at the slit. This harmonic mixed boundary-value problem is solved by a dual-series method.

No rigorous justification of the analysis is given, but an error bound is obtained.   A. R. Mitchell, United Kingdom

Fig. 6.3. Examples of critical abstracts. *Source. Applied Mechanics Reviews,* **21,** No. 12, December 1968, published by American Society of Mechanical Engineers, New York, N.Y., 10017. Reprinted with permission.

8835. Perkitny, T., and Nowicki, E., Investigations on the warping forces exerted by drying wood (in German), *Holz als Roh-und Werkstoff* **25**, 6, 217–225 (June 1967).

This rather wordy paper discusses the results of a series of experiments to determine the forces necessary to prevent the deformation of wood specimens during the drying process. Small-beam specimens with different grain orientation are constrained on three supports, with the measured center reaction considered as the characteristic holding force. Test results indicate considerable variation of the holding force as drying proceeds, with two peak values and, in some cases, actual reversal of the force.

This paper suffers severely from the purely empirical approach to the problem, though it appears to this reviewer that an analytical treatment, considering differential drying shrinkage occurring simultaneously with relaxation due to creep, would be entirely feasible.   K. Gerstle, USA

8959. John, K. W. Graphical stability analysis of slopes in jointed rock, *Journal of the Soil Mechanics and Foundations Division, Proceedings of the American Society of Civil Engineers* **94**, SM 2 497–526 (Mar. 1968).

A graphical method to evaluate the stability of a jointed rock mass bounded by planes of weakness along the joints and by free surfaces is presented.

The analysis, comprehensive of a geological and of an engineering phase, is carried out in three principal steps. A digest of geological data is first necessary to represent in a graphical hemispheric projection the spatial orientation of planes and lines and their relations to each other. Then a geotechnical model is designed, and the directions of potential sliding movements of the rock mass are determined. Finally, the factor of safety against such movements is evaluated by means of spatial polygons of forces, assuming frictional contacts along the joints. Blanks of the principal forms proposed for the analysis are given in appendix.

Paper illustrates very well a type of collaboration in series between geologists and engineers and encourages the extension of applied mechanics in a very promising field.
                                              R. Jappelli, Italy

Fig. 6.3. *(Continued)*

9082. du Prey, E. L., Method of interpreting the sessile drop for measuring interfacial tension and contact angle (in French), *Revue de l'Institut Francais du Petrole et Annales des Combustibles Liquides* **23**, 3, 365–373 (Mar. 1968).

Study of the forms of sessile drops has an old history in scientific literature, both for its own sake and—as in the present study—as a means of elucidating the basic properties of the liquid(s) studied, such as surface or interfacial tension, density, or viscosity. Present study is a contribution to this area of inquiry, and among its references is an early treatise [Bashforth and Adams, *An attempt to test the theories of capillary action by comparing the theoretical and measured forms of drops of fluid*, Cambridge University Press (1883)]. Paper describes a new method of interpreting the sessile drop for measuring the interfacial tension and the contact angle of two liquids. The method consists in measuring (by microscope) the diameter of the drop and the contact angle between the supporting surface and the interior contour of the drop. With the known density of the liquid and the (unknown) interfacial tension, a dimensionless number is formed. Author constructs a chart of two intersecting family of curves on a coordinate system with the contact angle as the abscissa and the dimensionless number as the ordinate. The point of intersection defines the interfacial tension. An accuracy of one degree for the angle of contact and 5% for the interfacial tension is claimed. Headings are "Advantages and inconveniences of the method of the sessile drop;" "Calculation of the drop profile" (given in mathematical detail); "Various methods of interpretation of the experimental profiles" (empirical formulas and use of the chart); "Method proposed in present study"; "Study of accuracy of the method of interpretation;" "Comparison of advantages and inconveniences of the present method of interpretation."

Paper is of interest to students of the physics of liquids and is worthy of study. With the antiquity of this important field, and the numerous, mostly piecemeal treatments of its plentiful facets at the hands of many worthy scientists [see Maxwell, "Capillary action" 9th ed. (1875) of *Encyclopedia Britannica*, in which about 30 previous researchers are named] the question arises: Would it not be proper and timely for some interested scientist in some appropriate institution to undertake a comprehensive and definitive treatment of the sessile drop, its theory, measurement, and conclusions drawn from it? Such a treatment should serve, hopefully for many years, as a reliable background and basis for future work involving interfacial tension.

K. J. De Juhasz, USA

Fig. 6.3. *(Continued)*

9163. **Murray, M. T., Propeller design and analysis by lifting surface theory,** *International Shipbuilding Progress* **14,** 160, 433–451 (Dec. 1967).

Author outlines a lifting-surface theory and the resulting computer program in use at the Admiralty Research Laboratory. The analysis generally follows present-day linearized theories [e.g., AMR **15**(1962), Rev. 4793; AMR **16**(1963), Rev. 4755; AMR **17** (1964), Revs. 377, 3447, and Rev. 3448; AMR **18**(1965), Revs. 437 and 6954] and applies to moderately-loaded propellers. In addition to the normal treatment of single propellers, author treats the case of contrarotating propellers, and the similar case of a rotor-stator combination. Both the design of propellers to meet specified performances and the calculation of the performance of given propellers are considered. Author has considered numerous aspects of propeller theory, but the discussion of either the theories or computer programs is not complete enough for use without redevelopment.

Reviewer believes that author should have stated certain restrictions for practical application of the computer program, such as (1) the desirability of making as much use as possible of lifting-line calculations, (2) the necessity of using the NACA $a = 0.8$ load distribution for the design problem, and (3) the difficulty of analyzing the performance of propellers with arbitrary chordwise load distributions. The last two restrictions arise from viscous effects.

W. B. Morgan, USA

Fig. 6.3. *(Continued)*

ABSTRACT NO. 068578   CA VOL. 72, NO. 14

KUNIN R, GUSTAFSON RL.   (ROHM AND HAAS CO., PHILADELPHIA, PA).

ION EXCHANGE.

IND. ENG. CHEM. VOL. 61, NO. 12, PP. 38-42, 1969.   (ASTM CODEN: IECHA)

INDEX TERMS: HYDROMETALLURGY MEMBRANE LIQ BIOCHEM PHARMACEUTICAL CATALYSIS WATER WASTE TREATMENT FOOD PROCESSING REVIEW

SEARCH TERMS PRESENT: WATER WASTE TREAT

ABSTRACT NO. 067982   CA VOL. 72, NO. 14

SYKORA S, KARASEK O, NAVRATIL B, MATULA M.

SOLVENTS REMOVAL FROM ELASTOMERS BY WASHING WITH WATER.

CZECH. PATENT NO. 130767 (CL. B 290),   15 JAN 1969, APPL. 02 FEB 1967; 3 PP.. (ASTM CODEN: CZXXA). ASSIGNEE: SYKORA, STANISLAV; KARASEK, OTAKAR; NAVRATIL, BOHUMIL; MATULA, MIROSLAV.

INDEX TERMS: SYKORA, STANISLAV; KARASEK, OTAKAR; NAVRATIL, BOHUMIL; MATULA, MIROSLAV POLYURETHANEUREAS POLYESTER URETHANES SOLVENT ACRYLONITRILE BUTADIENE COPOLYMERS PURIFN

SEARCH TERMS PRESENT: WATER PURIF

Fig. 6.4. Examples of titles "enriched" by use of index terms. *Source. CA— Condensates,* **72,** No. 14, 1970. Published by permission of the Chemical Abstracts Service of the American Chemical Society.

interest of speed but ordinarily with the understanding that a full abstract is available elsewhere or will follow later with all the details.

The mini-abstract and related approaches offer the potential advantage to the user of a rapid one-step process. The user can have before him on the same page both index entry and mini-abstract. This is in lieu of the usual two-step process in which the user first looks through index entries and then looks at abstracts of likely interest, a process which involves use of separate pages and possibly of separate volumes.

6.6. STATISTICAL OR NUMERICAL ABSTRACTS

In this kind of abstract, data are presented in tabular or numerical form. It is particularly suited to projections of use and other market trends. For example, see Fig. 6.5 (see p. 74).

This technique is also applicable to scientific papers in which the author summarizes results in a data table, which can be reproduced with the citation to form an abstract. Advantages of this kind of abstract are that it is concise and easily read. Its advocates also claim that because it is strictly statistical or numerical in nature it is likely to be more objective than the corresponding verbal abstract.

6.7. AUTHOR ABSTRACTS

Author abstracts (abstracts written by the author of the original document) have become very important in recent years, especially so now that they are increasingly being used by published abstracting services. As we have noted, author abstracts are now very often an important part of the refereeing process prior to publication, and, if so, they are subject to the same careful scrutiny given to the full text.

The abstractor may decide that the author abstract is completely adequate or needs only minor editing, thus saving time and effort. In other cases, however, the author abstract may be unsatisfactory; or there may be no author abstract at all—this often occurs in documents not published by one of the major technical societies.

The abstractor should be on the alert for defects in the author abstract. In the first place, for example, the author abstract may be poorly written. It can be grammatically faulty; it may include either too much data or not enough. It may not be backed up by facts in the full text. It may reflect only what the author considers important and omit material of possible interest to others. Defects such as these can prompt an abstractor either to modify the author abstract sharply or to scrap it entirely and write his own. In this event, the abstractor may still be able to save time and effort by using, as appropriate, direct excerpts from the full text of the original document. Also, there is almost never a need to redraw graphic material from the original. If the abstractor decides to incorporate graphics (such as chemical structures) from the original into his abstract, he can often realize substantial savings in time.

Many scientific journals now include multilingual author abstracts. For example, many German journals include abstracts in English and other languages, in addition to the original German abstracts. In his handling of author abstracts, the abstractor should pay special attention to abstracts of the multilingual species just described. This is because in the process of translating from one language into another, some of the precise meaning may be lost, there may be awkward sentence construction, or there may be errors in translation. The situation may call for editing to satisfy the needs of an English language audience or to correct errors.

6.8. DESCRIBING KIND AND CONTENT OF ABSTRACT

Abstractors can make their products more useful to readers by describing the abstracts they have written. This can be a further aid to the reader in deciding whether to consult the original.

For example, if no additional substantive information is found in the article, a phrase such as the following can be placed at the end of the abstract:

(no more info. in article)

Or if there is more substantive information in the original article:

(more)

Also, the abstractor may wish to make note at the end of the abstract of special features of the original document such as:

Photographs
Diagrams
Graphs
Tables
Infrared spectra

As an aid to readers desiring to scan abstracts quickly, the principal orientation or thrust of an article that has been abstracted can be specified in words following the citation. For example:

(Manufacturing Process Technology)
(Lab Scale Process)
(Theoretical Aspects)
(Applications, Uses)
(Marketing Data)

If this description is underlined or printed in boldface type, it enables the reader to bypass abstracts if he so desires. In this way, a scientist interested only in the process technology of a product can skip abstracts that cover documents on other aspects.

Fig. 6.5. Example of statistical abstracts. *Source.* Predicasts, Inc. Reprinted with permission.

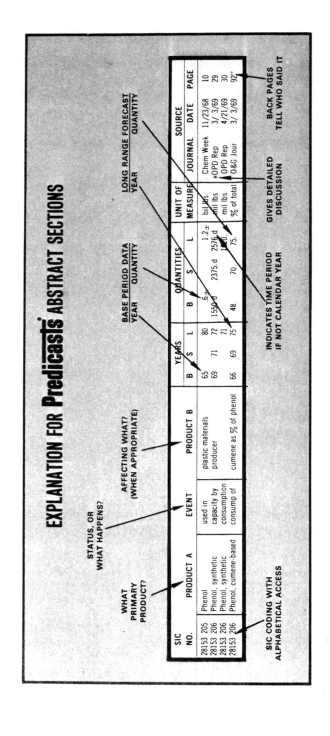

DETAILED PRODUCT PAGES

Coverage: (1) Forecasts on industries and 10,000 specific products for the United States, total North America and total world; (2) references to all good market articles containing detailed producer capacities, end use market distribution, market channel distribution, and historical trend data.

KEY TO COLUMN HEADINGS & SYMBOLS

SIC No.—PREDICASTS uses a modified Standard Industrial Classification for coding products and industries. With these codes, information can be correlated with data from government agencies and other sources. For an alphabetical cross-reference to the SIC, see Appendix.

Product A—The product or industry to which the statistic refers. Refers to the US unless otherwise stated.

Event—The transaction to which the statistic refers (production, consumption, etc.) or to status (capacity, in use, etc.)

Product B—The other product or item involved in the transaction. This space is also used to clarify information of other columns. When so used, parentheses ( ) are placed around the information.

Years—Base period, short range period, and long range period are indicated by column headings B, S, and L, respectively.

Quantities—The entries in the columns BSL refer to the quantities produced, consumed, etc. in the corresponding BSL years. Symbols used indicate: + (more than), − (less than), ± (about or average). When the time period ends in other than the normal calendar year the following symbols are used: a (1st qtr), b (2nd qtr), c (3rd qtr), d (4th qtr), f (fiscal year), k (model year), and z (21st century). For 'point-in-time' entries (capacity, population, etc), the symbols refer to a specific date occuring within the indicated time period. For capacity, 1968d indicates a quantity achieved during the 4th quarter of 1968. For production, 1968c indicates production for the year ending in the 3rd quarter. A '?' in the quantity column indicates potential rather than a probable realization of the forecast.

Unit of Measure—States the quantitative measure (tons, lbs., $, etc.) or relative measure (index, growth/yr., etc.).

Source—The abbreviation, date, and page number of the publication from which the statistic was abstracted. See the Appendix for detailed information on sources. The # symbol indicates a key market data article. The '' symbol after the page number indicates that the person or organization which made the forecast is listed in the Forecast Quote Pages. Arrangement of these pages is alphabetical by journal, then by date, and finally by page number. Small letters are used to distinguish several forecasts appearing on the same journal page.

Fig. 6.5. (*Continued*)

Another way to enable a reader to scan quickly is to highlight abstracts of especially significant documents with a symbol such as * or #.

The reader of this book can probably think of additional descriptions and other aids that might be useful in specific situations.

chapter 7
Next Step: Abstract Proper

When the abstractor is preparing to begin writing the text of the abstract proper, he should first scan the document at least once to get some idea of the subject matter, length, and special features, such as tables, illustrations, and lists of references. He is then ready for a second more careful reading with emphasis on the author abstract (if any), the first and last paragraphs, and key sections in the original document with headings such as "Introduction," "Purpose," "Conclusions," "Summary," and "Recommendations." These paragraphs usually contain the gist of what the author considers to be important and hence are important to the abstractor in identifying document content. Some abstractors find it helpful to underline key phrases and sentences during their readings of the document.

If there is no author abstract, or if it is unsatisfactory, the abstractor usually works with the full text of the original document, and especially with the key sections just mentioned. The abstractor can attempt to paraphrase in concise form what the author of the original document has said. But he will often want to retain as much as possible of the original emphasis and terminology for the sake of accuracy. He may want to use brief direct excerpts to prevent changing meanings because of any subjective leanings he may have or because of excessive zeal for compactness.

A beginning abstractor will probably want to write a rough draft first and from this prepare a final copy. Some experienced abstractors proceed directly to the final copy, but even experienced workers may prefer to write a draft. This is a matter of personal working habits and preference. It also leads to a fundamental principle of good abstracting:

The abstractor should try to work and write in the style and manner with which he feels most comfortable and efficient, consistent with the ground rules laid down by his organization (or by himself in the case of a one-man organization).

7.1. ABSTRACT CONTENT AND LENGTH

The perceptive abstract reader expects to find the following kinds of information about the original document in the informative abstract (not necessarily in this sequence):

1. Purpose.
2. Results.
3. Methods, techniques, instruments.
4. Conclusions.

Factors that can specifically affect abstract content and length are listed in the sections which follow.

7.1.1.   Client Needs

This is the most important factor. Users in some organizations or fields may dislike intervention (such as lengthy abstracts) between themselves and original documents. Some are quite satisfied with only a good citation or a brief, indicative abstract. Others need lengthy, informative abstracts. Among the variables which can affect client needs are the fields in which they work and the ease of access to a library that contains the original documents. Users with ready access to strong libraries may require less lengthy abstracts. Ascertaining client needs is not easy, but the effort to learn something about these needs is well worth it.

7.1.2.   Scope and Nature of the Original Document and the Publication in Which It Appears

Papers of broad scope or those covering major original work usually should get fuller treatment (more lengthy abstracts) than papers of more limited originality or scope. Unfortunately, not all abstractors are qualified to make this kind of judgment.

7.1.3.   Field (this is related to Section 7.1.1.)

In dynamic, rapidly moving fields or in those of immediate commercial impact, time limitations may compel the abstractor to write briefer and relatively unsophisticated abstracts. Also, in different fields, different aspects may be emphasized and differing degrees of detail are called for. For example, an abstract of a paper on new methods for the synthesis of many organic compounds in a related series often is abstracted in great detail and can run many hundreds of words. The reader (usually an organic chemist) has come to expect this. But in this case length can be controlled somewhat by giving a single example from the series and stating that other compounds were similarly prepared. The trend is now toward somewhat shorter abstracts in organic chemistry. But scientists engaged in use, evaluation, or compilation of physical constants may like to see as much detail as possible in an abstract; in this case, length control is difficult.

7.1.4. Abstractors—Numbers and Kinds

The more the total hours available for abstracting, the more the opportunity for writing more informative abstracts. The skill and scientific expertise of the abstractor(s) also affects the length and kind of abstract that can be written.

7.1.5. Budget for Abstracting Operations (this is related to Section 7.1.4.)

The budget has a significant impact on the amount of time and effort that can be spent in the writing of abstracts.

7.1.6. Language of Original Document

Everything else being equal, abstractors usually try to make abstracts of foreign language documents more complete in detail than for English language documents. One reason is that many foreign language documents are difficult, and sometimes almost impossible to obtain, especially for users far from a major research library. Moreover, once the document is obtained, if the user cannot read the foreign language, he often needs to have someone else translate for him. This can be a costly and time-consuming process. A good abstract with sufficient detail can enable the user to decide whether to obtain the full original or to request a translation.

7.1.7. Availability of Original Document

If a document is widely read and circulated, hence readily accessible, some abstractors may choose to abstract more briefly than for less available documents, everything else being equal.

7.1.8. Intended Use of the Abstract

An abstract intended for quick alerting purposes would be briefer (80 to 100 words) and less informative. On the other hand, abstracts intended for more permanent reference use might be longer (400 to 500 words) and more informative. Use of diagrams or drawings would tend to cut down on the number of words in either case.

7.1.9.   Subsequent Use of the Abstract Output as a Physical Unit

In some situations abstracts must be written to fit within a prescribed space, or onto a specific form (like a 3 $\times$ 5 card or a punched card). This could be because either the generators or the users of the abstracts intend that the output physically fit into some kind of filing system for ready access by users. This kind of use imposes abstract length limitations.

7.1.10.   Guidelines

In some cases standards of length are set by management or the abstractor in the guidelines. For example, an abstract length of "preferably less than 3 percent of the original, usually 150 to 300 words," is suggested in *Guide to the Preparation of Abstracts* issued by *Biological Abstracts.*

7.2.   WRITING ABSTRACTS CONCISELY

The good abstractor is fully aware that readers of his abstracts are likely to be busy people. Therefore, he strives for abstracts that are concise and succinct (but complete). He does not waste words. He avoids repetitive and meaningless expressions. He knows that superlatives and other adjectives are not usually necessary. As explained below, he makes skillful use of jargon, abbreviations, symbols, and trade names.

He also saves space by spelling out lengthy names just once and referring to them thereafter by a numerical designation such as a Roman numeral. This practice also saves time and avoids typographic errors. For example, consider the following fictitious sample:

> The cost factors involved in the manufacture of product WXYZ(I) are affected by plant location. Because of closeness to raw materials and markets, location ABC is regarded as the most economic location for the manufacture of (I). The demand for (I) is expected to remain strong for the next 10 years.

The reader of this book may want to apply his own ingenuity to making the above abstract more concise.

Another way to save space is to avoid repeating the words of the title. For example, for a patent titled:

Manufacture of Products A, B, C, D, and E

the abstract can begin:

The title products were made using process XYZ. . . .

Although some abstractors frown on the use of incomplete sentences or telegraphic phrases, when skillfully used, these sentences or phrases can save space. There is nothing basically wrong with incomplete sentences if the abstractor gets the "message" across clearly. For example:

New high yield process for mfr. of product 123 developed by XYZ Co. involves use of novel precious metal catalyst (not identified) under conventional conditions. Yields of product exceed 70% of theoretical. Product quality considered superior to normal commercial limits listed in Standard 1234.

It can be seen from the example that the use of telegraphic phrases or incomplete sentences is, in a way, like writing descriptive newspaper headlines.

7.3 LANGUAGE SHORTCUTS

These include trade names, abbreviations, symbols, and jargon. Such language shortcuts make for faster writing, reading and typing, and for a compact abstract. The disadvantages, however, may outweigh the benefits for the abstract user. The principal disadvantage to the user is that excessive use of language shortcuts can lead to confusion as to precisely what is meant. In the exacting world of science, misunderstanding of word meanings is an ever present threat. Discretion is needed in the use of language shortcuts.

A misunderstanding is especially likely to happen with the passage of time as people forget the exact meaning of language shortcuts, and it is also especially likely to happen in interdisciplinary areas. Consider this oversimplified example: The meaning of DDT is very well known today (1970), but will the meaning be equally unambiguous in the year 2000? To cite another example, the word "cell" means one thing to the industrial chemist and another thing to the biologist. To the industrial chemist the word "cell" usually denotes an electrochemical device or system. To the biologist the word "cell" denotes a small unit of living matter.

7.3.1. Trade Names and Codes

Most trade names need identification on a regular basis in each abstract in which they are used. There are so many trade names it is impossible to identify all of these in any kind of convenient list. In addition to improving reader understanding of the abstract proper, trade-name identification is useful in the subsequent *indexing* of the abstract. Unless abstracts are *indexed* under both trade name and meaning of the trade name, the abstracts are likely to be missed and "lost" by index users. For these reasons, it is desirable for the abstractor to identify the meanings of trade names as they are used within the text of the abstract. This may require extra effort on the part of the abstractor, and he may not always succeed in getting identification, but it is worth a good try. For example, here is how a trade name might be identified by an abstractor:

The Omadines® (1-hydroxypyridine-2-thione derivatives), biocides made by Olin Corporation, were studied in connection with . . . .

See also Fig. 7.1.

97. GAS CHROMATOGRAPHY OF ABATE. Ralph W. Jennings and August Curley, National Communicable Disease Center, Toxicology Laboratory, Pesticides Program, Atlanta, Ga. 30333

The gas-liquid chromatography of Abate (O,O,O'O'-tetramethyl-O,O'-thiodi-p-phenylene phosphoro-thioate) has been attempted by several researchers. Success is reported by Wright et al (J. Agr. Food Chem. 15:1038-309, 1967) on Dow 11 but under rather rigorous conditions, e.g. column temperature of 250°C. Successful detection of Abate has been achieved on 3% OF-1 and 3% SE-30 columns at temperatures ranging from 150°C to 180°C. The detection system was the flame photometric detector with 526 mu filter. The observed retention times indicate that gas chromatographers may have been losing Abate in the solvent peak at higher temperatures.

Fig. 7.1. Defining the tradename within the abstract. *Source. Abstracts of Papers,* ACS National Meeting, Atlantic City, N.J., September 1968. Published by the American Chemical Society. Reprinted with permission.

Codes (such as "XYZ/2345") are totally unintelligible to most scientists. The abstractor should try to give some kind of definition within the abstract if at all possible; even a definition of the broad class (e.g., "aromatic fluorine compounds") can be helpful in such cases.

If, for some reason, the trade name cannot be identified, the abstractor should so indicate by such means as placing in parentheses thereafter the words "not identified by author" or "unidentified trade name."

7.3.2. Abbreviations, Symbols

Abbreviations and symbols save space and time. For example, "MLD" is shorter than "minimum lethal dose" and "%" is shorter than "percent." To ensure proper use of abbreviations in an abstracting service, both abstractors and abstract users should have a standard and current list of abbreviations made available to them by the editors of the service. Examples of such a list are shown in Figs. 7.2 and 7.3.

ACCEPTABLE ABSTRACT ABBREVIATIONS

A (or amp)	ampere
A	Angstrom
AC	alternating current
ACTH	adrenocorticotropic hormone
A.D. (with dates)	anno Domini
ad lib.	ad libitum
ADP, dADP	adenosine diphosphate, deoxyadenosine diphosphate
ADPase	adenosine diphosphatase
a.m. (with time)	ante meridiem
amp (or A)	ampere
AMP, dAMP	adenosine monophosphate, deoxyadenosine monophosphate
AMPase	adenosine monophosphatase
atm	atmosphere
ATP, dATP	adenosine triphosphate, deoxyadenosine triphosphate
ATPase	adenosine triphosphatase
auct.	auctorum (of authors; taxonomy only)
B.C. (with dates)	before Christ
BCG	bacille Calmette Guerin
BHC	benzene hexachloride; hexachloro-cyclohexane
BMR	basal metabolic rate
BTU	British thermal unit
bu	bushel
c	centi- (prefix, 10^{-2})
C (or coul)	coulomb
°C	degrees Celsius (Centigrade)
cal	gram calorie (use kcal for kilogram calorie)
cc (or cm³)	cubic centimeter
CDP, dCDP	cytidine diphosphate, deoxycytidine diphosphate
Ci	curie
cm	centimeter
CMP, dCMP	cytidine monophosphate, deoxycytidine monophosphate
CNS	central nervous system
CoA	coenzyme A
comb. nov.	combinatio novum (new combination; taxonomy only)
coul (or C)	coulomb
cps	cycles per second
CTP, dCTP	cytidine triphosphate, deoxycytidine triphosphate
cwt	hundred weight
cv.	cultivar (following a specific name only)
db	decibel
2,4-D	2,4-dichlorophenoxyacetic acid
DC	direct current
DDT	dichlorodiphenyltrichloroethane
DEAE cellulose	diethylaminoethyl cellulose
DFP	diisopropylphosphorofluoridate

Fig. 7.2. Abbreviations. *Source. Biological Abstracts* (BIOSIS). Reprinted with permission.

This is a continuation of Fig. 3.1.

DNA	deoxyribonucleic acid
DNase	deoxyribonuclease
DOPA	dihydroxyphenylalanine
DOPAMINE	dihydroxyphenethylamine
doz	dozen
DPN (or DPN+)	diphosphopyridine nucleotide (NAD)
DPNH	reduced diphosphopyridine nucleotide (NADH)
dr	dram
E	east
ECG (preferred to EKG)	electrocardiogram
ECHO	enteric cytopathogenic human orphan (virus)
EDTA	ethylenediaminetetraacetate
EEG	electroencephalogram
e.g.	for example
emend.	emendation, emended (taxonomy only)
EPR	electron paramagnetic resonance
eV	electron volt
etc.	et cetera
f.	forma (form; following a specific name only)
°F	degrees Fahrenheit
F_1, F_2	first filial generation, second filial generation
FAD	flavine-adenine dinucleotide
fam. nov.	famila nova (new family; following a familial name only)
FAO	Food and Agriculture Organization
fl oz	fluid ounce
FMN	flavine mononucleotide
ft	foot, feet
ft-c	foot-candle
g	gram
G	giga (prefix, 10^9)
GA	gibberellic acid
gal	gallon
gen. nov.	genus novum (new genus, following a generic name only)
GDP, dGDP	guanosine diphosphate, deoxyguanosine diphosphate
GMP, dGMP	guanosine monophosphate, deoxyguanosine monophosphate
GTP, dGTP	guanosine triphosphate, deoxyguanosine triphosphate
ha	hectare
Hb	hemoglobin
HbO_2	oxygenated hemoglobin
hp	horsepower
hr (or h)	hour
Hz	hertz
IAA	indoleacetic acid
IDP, dIDP	inosine diphosphate, deoxyinosine diphosphate
i.e.	that is
i.m.	intramuscular(ly)

Fig. 7.2. *(Continued)*

IMP, dIMP	inosine monophosphate, deoxyinosine monophosphate
in.	inch
inc. sed.	incertae sedis (uncertain position; taxonomy only)
i.p.	intraperitoneal(ly)
IR	infra-red
ITP, dITP	inosine triphosphate, deoxyinosine triphosphate
IU	international units
i.v.	intravenous(ly)
k	kilo (prefix, 10^3)
°K	degrees Kelvin
kcal	kilocalorie
keV	kiloelectron volt
kg	kilogram
kV	kilovolt
l	liter (use only when meaning is clear)
lb	pound
LD_{50}	lethal dose, median
m	meter; milli (prefix, 10^{-3})
m.	morpha (form; following a specific name only)
M	mega (prefix, 10^6)
$\underline{M}$	molar (concentration)
$\underline{\mu}$	micron; micro (prefix, 10^{-6})
μeq	microequivalent
μg	microgram
mA	milliampere
meq	milliequivalent
MeV	million electron volt
mg	milligram
mi	mile
min	minute
ml	milliliter
MLD	minimum lethal dose
mM	millimolar (concentration)
mm	millimeter
mmole	millimole (mass)
μmole	micromole (mass)
mo.	month
months, names of	Jan., Feb., Aug., Sept., Oct., Nov., Dec.
mph	miles per hour
mRNA	messenger ribonucleic acid
mV	millivolt
n	nano (prefix, 10^{-9})
$\underline{N}$	normal (concentration)
$\overline{N}$, NE, NW	North, Northeast, Northwest
NAD (or NAD+)	nicotinamide adenine dinucleotide
NADH	reduced nicotinamide adenine dinucleotide
NADP (or NADP+)	nicotinamide adenine dinucleotide phosphate
NADPH	reduced nicotinamide adenine dinucleotide phosphate
NMR	nuclear magnetic resonance
nom. dub.	nomen dubium (doubtful name; taxonomy only)

Fig. 7.2.   *(Continued)*

nom. cons.	nomen conservandum (retained name; taxonomy only)
nom. nov.	nomen novum (new name; taxonomy only)
nom. nud.	nomen nudum (invalid name; taxonomy only)
nom. rej.	nomen rejiciendum (rejected name; taxonomy only)
oz	ounce
p	pico (prefix, 10^{-12})
P	probability
pH	log of the reciprocal of the hydrogen-ion concentration
p.m. (with time)	post-meridiem
PPLO	pleuropneumonia-like organisms
ppb	parts per billion
ppm	parts per million
preocc.	preoccupied (taxonomy only)
pt	pint
psi (or lb/in^2)	pounds per square inch
Q_{10}	the change in the rate of a process when temperature is increased 10°C
R	roentgen
rad	radiation, ionizing absorbed dose (100ergs/g of irradiated material)
rem	Roentgen equivalent man
RES	reticuloendothelial system
R_f	retardation factor (distance the unknown has traveled relative to the solvent front in chromatography)
RNA	ribonucleic acid
RNase	ribonuclease
rpm	revolutions per minute
RQ	respiratory quotient
S, SE, SW	South, Southeast, Southwest
s.c.	subcutaneous(ly)
SD	standard deviation
SE	standard error
sec (or s)	second
sect.	section (taxonomy only)
s.l.	sensu lato (broad sense; taxonomy only)
sp., spp.	species (following a generic name or numeral only)
sp. nov.	species novum (new species; following a new specific name only)
sRNA	soluble RNA
s.s.	sensu stricto (restricted sense; taxonomy only)
ssp., sspp.	subspecies (following a specific name only)
ssp. nov.	subspecies novum (new species; following a subspecific name only)

Fig. 7.2.   *(Continued)*

TDP, dTDP	thymidine diphosphate, deoxythymidine diphosphate
TEAE cellulose	triethylaminoethyl cellulose
THIOTEPA	triethylenethiophosphoramide
TMP, dTMP	thymidine monophosphate, deoxythymidine monophosphate
TPN (or TPN+)	triphosphopyridine nucleotide
TPNH	reduced triphosphopyridine nucleotide
tRNA	transfer ribonucleic acid
TTP, dTTP	thymidine triphosphate, deoxthymidine triphosphate
UDP, dUDP	uridine diphosphate, deoxyuridine diphosphate
UMP, dUMP	uridine monophosphate, deoxyuridine monophosphate
UNESCO	United Nations Educational, Scientific and Cultural Organization
USA	United States of America
U.S.P.	United States Pharmacopeia
USSR	Union of Soviet Socialist Republics
UTP, dUTP	uridine triphosphate, deoxyuridine triphosphate
UV	ultraviolet
V	volt
var.	varietas (variety; following a specific name only)
var. nov.	varietas nova (new variety; following a new varietal name only)
vol.	volume (book)
vs.	versus
W	watt; West
WHO	World Health Organization
wk	week
XDP, dXDP	xanthosine diphosphate, deoxyxanthosine diphosphate
XMP, dXMP	xanthosine monophosphate, deoxyxanthosine monophosphate
XTP, dXTP	xanthosine triphosphate, deoxyxanthosine triphosphate
yd	yard
yr	year
°	degrees
Ω	ohm
/	per
%	percent (following a number)
o/oo	per mille (per thousand; following a number)
*	probability level = .05
**	probability level = .01

Fig. 7.2. *(Continued)*

Fig. 7.3. Abbreviations. *Source. Directions for Abstractors,* Chemical Abstracts Service, American Chemical Society. Reprinted with permission.

ABBREVIATIONS AND SYMBOLS USED IN CHEMICAL ABSTRACTS

A	ampere
Å	angstrom unit
abs.	absolute
abstr.	abstract
Ac	acetyl (CH_3CO, not CH_3COO)
a.c.	alternating current
ACTH	adrenocorticotropin
addn.	addition
addnl.	additional(ly)
ADP	adenosine 5'-diphosphate
alc.	alcohol, alcoholic
alk.	alkaline (not alkali)
alky.	alkalinity
AMP	adenosine 5'-monophosphate
amt.	amount
amu	atomic mass unit
anal.	analysis, analytical(ly)
anhyd.	anhydrous
AO	atomic orbital
app.	apparatus
approx.	approximate(ly)
approxn.	approximation
aq.	aqueous
assoc.	associate
assocd.	associated
assocg.	associating
assocn.	association
at.	atomic (not atom)
atm	atmosphere (the unit)
atm.	atmosphere, atmospheric
ATP	adenosine 5'-triphosphate
ATPase	adenosinetriphosphatase
av.	average
b.	(followed by a figure denoting temperature) boils at, boiling at (similarly b_{13}, at 13 mm pressure)
bcc.	body centered cubic
BeV or GeV	billion electron volts
BOD	biological oxygen demand
μB	Bohr Magneton
b.p.	boiling point

crystg.	crystallizing
crystn.	crystallization
CTP	cytidine 5'-triphosphate
d-	deci- (as a prefix, e.g., dl)
d.	density (d^{13}, density at 13° referred to water at 4°; d^{20}_{20}, at 20° referred to water at the same temperature)
D	debye unit
d.c.	direct current
DEAE-cellulose	diethyl-aminoethyl cellulose
decompd.	decomposed
decompg.	decomposing
decompn.	decomposition
deriv.	derivative
det.	determine
detd.	determined
detg.	determining
detn.	determination
diam.	diameter
dil.	dilute
dild.	diluted
dilg.	diluting
diln.	dilution
dissoc.	dissociate
dissocd.	dissociated
dissocg.	dissociating
dissocn.	dissociation
distd.	distilled
distg.	distilling
distn.	distillation
DNA	deoxyribonucleic acid
DNase	deoxyribonuclease
d.p.	degree of polymerization
dpm	disintegrations per minute
DPN	diphosphopyridine nucleotide (NAD)
DPNH	reduced DPN
DTA	differential thermal analysis
ED	effective dose

G-	giga-(10^9)
g	gram
g	gravitation constant
GDP	guanosine 5'-diphosphate
GMP	guanosine 5'-monophosphate
GTP	guanosine 5'-triphosphate
H	henry
ha	hectare
Hb	hemoglobin
hr	hour
Hz	hertz (cycles/sec)
ICSH	interstitial cell-stimulating hormone
ID	infective dose
IDP	inosine 5'-diphosphate
i.m.	intramuscular
IMP	inosine 5'-monophosphate
inorg.	inorganic
insol.	insoluble
i.p.	intraperitoneal
ir	infrared
irradn.	irradiation
ITP	inosine 5'-triphosphate
IU	International Unit
i.v.	intravenous
J	joule
k-	kilo- (as a prefix, e.g., kg)
l.	liter
lab.	laboratory
LCAO	linear combination of atomic orbitals
LD	lethal dose
LH	luteinizing hormone
liq.	liquid
lm	lumen
lx	lux
m-	milli- (as a prefix, e.g., mm)
m.	meter
m.	melts at, melting at
m	molal
M-	mega- (10^6)
M	molar

phys.	physical(ly)
PMR	proton magnetic resonance
polymd.	polymerized
polymg.	polymerizing
polymn.	polymerization
pos.	positive(ly)
powd.	powdered
ppb	parts per billion
ppm	parts per million
ppt.	precipitate
pptd.	precipitated
pptg.	precipitating
pptn.	precipitation
Pr	propyl (normal)
prep.	prepare
prepd.	prepared
prepg.	preparing
prepn.	preparation
psi	pounds per square inch
psia	pounds per square inch absolute
psig	pounds per square inch gage
py	pyridine (used in Werner complexes only)
qual.	qualitative(ly)
quant.	quantitative(ly)
R	roentgen
redn.	reduction
ref.	reference
rem	roentgen equivalent man
rep	roentgen equivalent physical
resp.	respective(ly)
RNA	ribonucleic acid
RNase	ribonuclease
rpm	revolutions per minute
RQ	respiratory quotient
sapon.	saponification
sapong.	saponifying
sapond.	saponified
sat.	saturate
satd.	saturated
satg.	saturating
satn.	saturation

Bz   benzoyl (C_6H_5CO, not $C_6H_5CH_2$)
c-   centi- (as a prefix, e.g., cm)
cal   calorie
calc. calculate
calcd. calculated
calcg. calculating
calcn. calculation
CD   circular dichroism
c.d. current density
CDP   cytidine 5′-diphosphate
chem. chemical(ly)
chromatog. chromatography
Ci   curie
clin. clinical(ly)
CM-cellulose carboxymethyl cellulose
CMP   cytidine 5′-monophosphate
CoA   coenzyme A
COD   chemical oxygen demand
coeff. coefficient
com. commercial
compd. compound
compn. composition
conc. concentrate
concd. concentrated
concg. concentrating
concn. concentration
cond. conductivity
const. constant
contg. containing
cor. corrected
CP   chemically pure
crit. critical
cryst. crystalline (not crystallize)
crystd. crystallized

emu   electromagnetic unit
en   ethylenediamine (used in Werner complexes only)
EPR   electron paramagnetic resonance
equil. equilibrium(s)
equiv. equivalent
esp. especially
ESR   electron spin resonance
est. estimate
estd. estimated
estg. estimating
estn. estimation
esu   electrostatic unit
Et   ethyl
eV   electron volt
evap. evaporate
evapd. evaporated
evapg. evaporating
evapn. evaporation
examd. examined
examg. examining
examn. examination
expt. experiment
exptl. experimental(ly)
ext. extract
extd. extracted
extg. extracting
extn. extraction
F   faraday, farad
FAD   flavine adenine dinucleotide
fcc. face centered cubic
FMN   flavine mononucleotide
f.p. freezing point
FSH   follicle-stimulating hormone
G   gauss

manufg. manufacturing
math. mathematical(ly)
max. maximum(s)
Me   methyl (not metal)
mech. mechanical
min   minute (time)
min. minimum(s)
misc. miscellaneous
mixt. mixture
MO   molecular orbital
mol. molecule, molecular (not mole)
m.p. melting point
μ   micron; also micro- (as a prefix, e.g., μl)
MSH   melanocyte-stimulating hormone
Mx   maxwell
n-   nano- (10^{-9})
n   refractive index (n_D^{20} for 20° and sodium D light)
N   newton
N   normal (as applied to concn.)
NAD   nicotinamide adenine dinucleotide (DPN)
NADH reduced NAD
NADP nicotinamide adenine dinucleotide phosphate (TPN)
NADPH reduced NADP
neg. negative(ly)
NMN   nicotinamide mononucleotide
NMR   nuclear magnetic resonance
no.   number
obsd. observed
Oe   oersted
ORD   optical rotatory dispersion
org. organic
oxidn. oxidation
P   poise
p-   pico- (10^{-12})
p.d. potential difference
Ph   phenyl

SCF   self-consistent field
sec   second (time unit only)
sec   secondary (with alkyl groups only)
sep. separate(ly)
sepd. separated
sepg. separating
sepn. separation
sol. soluble
soln. solution
soly. solubility
sp.   specific (used only to qualify physical constant)
sp. gr. specific gravity
sr   steradian
St   Stokes
std. standard
sym. symmetrical(ly)
T-   tera- (10^{12})
TEAE-cellulose triethylaminoethyl cellulose
tech. technical
temp. temperature
tert tertiary (with alkyl groups only)
titrn. titration
TPN   triphosphopyridine nucleotide (NADP)
TPNH reduced TPN
Tris tris(hydroxymethyl)aminomethane
TSH   thyroid-stimulating hormone
UDP   uridine 5′-diphosphate
UMP   uridine 5′-monophosphate
USP   United States Pharmacopeia
UTP   uridine 5′-triphosphate
uv   ultraviolet
V   volt
vol. volume (not volatile)
W   watt
wt.   weight

Plurals of noun abbreviations are formed by adding "s" to the singular abbreviation except when a single abbreviation is designated to show both the singular and plural forms. Words formed by adding prefixes to words normally abbreviated are also abbreviated, as microchem. for microchemical. Other well established abbreviations, as etc., i.e., e.g., and abbreviations for English units of weight and measure, are also used. Unit abbreviations signify both singular and plural forms. Words ending in -ology or -ological(ly) are abbreviated -ol., e.g., geol. for geology.

Some abstractors (and abstract users), however, question the value of abbreviations. It is sometimes faster to spell out (or understand) a word in full than to take the time to look it up in a standard list. Also, the use of the full, unabbreviated word eliminates any chance of ambiguity and misunderstanding. For example, does the abbreviation "min." stand for minute, minimum, or minimal? The answer could depend on the abstract user and his interpretation, which might or might not be the same as that intended by the abstractor. It is important that the abstractor be aware of the kind of pitfall just mentioned.

Many words can be safely abbreviated or put into symbol or acronym form with almost no chance of misunderstanding. For example:

manufacture	mfr
degrees Centigrade	°C
Atomic Energy Commission	AEC
National Aeronautics and Space Administration	NASA

Some abbreviations can be identified within the text of the abstract. When an abbreviation is used for the first time within the text of an individual abstract, it is defined. The abbreviation can be used safely thereafter within that same individual abstract. For example:

PVC (polyvinyl chloride) is used extensively in construction. One of the principal advantages of PVC is its versatility. . . .

or

Toluene diisocyanate (TDI) was manufactured by process ABC. TDI is used in the manufacture of plastics such as. . . .

See also Fig. 7.4.

39. THE VALUE OF WOOD-DERIVED PRODUCTS IN RUMINANT NUTRITION. W. H. Pfander, Animal Husbandry Department, University of Missouri, Columbia, Mo. 65201

Wood chips subjected to high pressure steam for two minutes and washed yielded washings, containing 4% dry matter, which were concentrated to 65% dry matter. This residue is designated as crude hemicellulose (HC). The HC may be neutralized (NHC). HC contains: carbohydrate, 53; acetyl as acetic acid, 12; uronic anhydride, 6; lignin, 8; ash, 3. The carbohydrate distribution is: glucose, 15; galactose, 6; mannose, 25; arabinose, 5; and xylose, 49. Palatability and digestibility trials with lambs showed that HC was similar to cane molasses. A 2x2x2 factorial design was used to estimate the energy value of HC. Variables were: (1) source of HC, Mississippi (L), or California redwood (U); (2) HC or NHC; and (3) ration level, 15 or 30%. The addition of 15 and 30% HC improved protein digestibility and increased feed intake. LHC contained approximately 2860 kcal/kg at 30% of the ration, UHC contained less energy than LHC. NLHC and NUHC contained approximately 2420 kcal/kg. Trial 2 used 0, 15 and 30% of spray dried HC in a basal ration containing 1.8% fiber, 14.25% crude protein and 67.75% nitrogen free extract. Dried HC as 15% of the ration contained 3100 kcal/kg; however, when fed at higher levels, digestibility was reduced. NLHC at 50% also reduced digestibility. The lowered energy of dried HC was confirmed in cattle.

Fig. 7.4. Defining abbreviations within abstracts. *Source. Abstracts of Papers,* ACS National Meeting, Atlantic City, N. J., September 1968. Published by the American Chemical Society, Washington, D.C. Reprinted with permission.

What it all boils down to is this: abbreviations and symbols that are commonly used and widely known need not be spelled out. The abstractor uses his judgment and experience to arrive at a decision. If in doubt, he spells it out in full or defines it. At all times he is aware that an abbreviation, though widely known and fully accepted today, may become an ambiguous or even meaningless abbreviation within just a few years.

7.3.3. Trade Jargon

The use of trade jargon (if it must be used) is best confined to two situations:

1. In-house bulletins.
2. Highly specialized services for a specialist clientele.

At all other times, the abstractor should use scientific terminology.

Probably the widest use of jargon in American life today is found in telecasts of professional sporting events. A large proportion of the words used by the announcers need very clear definition for those "not in the know." Each technology also has its own jargon which also needs an explanation for those not in the field. The field of plastics is an example of a technology in which the use of jargon is especially common. The passage of time, however, can cloud the meaning of jargon (just as it can with abbreviations) even to a specialized technical audience. This important caution should be weighed against the possible savings in words and time and the jaunty, spicy flavor that jargon brings to almost any kind of writing.

7.4. RESPONSIBILITY AND CREDIT

We have indicated earlier that the abstractor can work more rapidly and more accurately if he uses direct excerpts from the original document or, better still, the full author abstract if such exists in acceptable form. We recommend that the abstractor use quotation marks for material that is partially excerpted and that he label full author abstracts as such. Not all abstracting service personnel may agree with this recommendation, but we believe that it is in the best interests of objectivity. This also helps clarify responsibility for the abstract. Examples of specifying credit are:

1. Author's abstract.
2. Translated author abstract.
3. From author's summary.
4. Translated from author's summary.

Credits such as the above (and the abstractor's name if he prepares or edits the abstract) belong at the end of the abstract.

7.5. ABSTRACT ARCHITECTURE

In a sense, a good abstractor is like a good radio, TV, or newspaper reporter. He puts first things first, with details or matters of lesser importance filled in later, towards the end of the abstract. The first or "lead" sentence is especially important and should give the gist or essence of the document being abstracted. One cogent reason that the lead sentence is so important is that some readers will go no further in their reading of an abstract.

As we have said, the good abstractor takes special care to avoid duplicating the words of the title. He also avoids such space-consuming beginning phrases as "This article deals with. . . ." Instead, he plunges into the essence of the document at hand with an opening sentence such as:

A new method for the preparation of product X in 80% yield is based on the reaction of A with B at 95°C at atmospheric pressure.

Another useful technique is to put the abstract first and to follow this by the citation. In an abstract so structured, the first thing the reader should see will be a highlighted first sentence. This is a very attractive method of presentation since many scientists are interested in abstract content first and the reference only secondarily, after they have established their initial interest. For example, see Fig. 7.5 (see pp. 94–95).

7.6. MAKING ABSTRACTS EASY TO READ AND USE

We have already noted some ways in which the abstractor can help the reader:

1. He puts the "essence" first.
2. He writes concisely.
3. He writes clearly and understandably.
4. He provides the full reference citation.

There are other techniques that can be used. The use of underlining, boldface, large type, or capital letters highlights lead sentences and important words or phrases in the body of the abstract. These are helpful signposts to the reader. For example, see Fig. 7.6 (see pp. 97–101).

Consider the following abstract as it would be read by specialists in the fictitious material XYZ:

A large number of materials of construction were evaluated for possible use in the ABC program. The materials studied were: several stainless steels, aluminum and copper alloys, titanium, beryllium, zirconium, <u>XYZ</u>, polyethylene, polyvinyl chloride (PVC), and polystyrene. The most desirable materials for this specific application were identified as PVC, <u>XYZ</u>, and polystyrene.

Note how the underlining enables the specialist in XYZ to locate quickly the references to XYZ in the abstract.

The abstractor should also bear in mind that, although capitalization and other "highlighting" techniques can be useful, this can be overdone to the extent that the wheat cannot be distinguished from the chaff. Underlining and other highlighting techniques may also cause problems with typography and with computer-based operations.

Effective use of graphics within abstracts is a word-and-work-saving measure for both the user and the writer of the abstract. For many scientists, the use of a sketch of mechanical apparatus, of a graph, or of a diagram of a chemical structure may often be more meaningful than a wordy description. Graphic material can, in many cases, be scissored or reproduced directly from the original document, and used in the abstract as such.

7.7. ORIENTING ABSTRACTS

Abstractors write oriented abstracts when they emphasize some aspects of documents and deemphasize other aspects. Often this is done to meet the needs of a specialized audience—but not always.

Some degree of orienting is natural, and indeed inevitable. After all, the abstractor is a product of years of education and of at least some experience in a specific discipline. His abstracts will almost inevitably reflect this background. A completely nonoriented abstract is too much to expect from anyone.

When the abstractor writes for an in-house bulletin, he is ordinarily under a definite *obligation* to orient his abstracts towards the needs of his organization. In his abstract, he should emphasize that part of the original document that is of special importance and interest to his colleagues. He will not ordinarily omit other important information but will give it much less emphasis. In-house abstractors can orient their abstracts easily since they usually know many of their "clients" (and their needs and interests) personally.

Specialized abstracting services aimed at external (public) but spe-

OXOCHEM STARTS BUILDING P.R. FACILITY. Construction is under way on Oxochem Enterprise's 250 million lb/yr Oxo-alcohol plant at Penuelas, P.R. The $25 million facility is to be completed in early 1970; Fluor Corp. Ltd. is the contractor for engineering, procurement, and construction. Plans include a synthesis gas unit, which will make hydrogen and synthesis gas from a feedstock obtained from nearby sources. Oxochem is a joint venture between subsidiaries of Commonwealth Oil Refining Co. Inc. and the Hatco Group of W. R. Grace & Co. (no more info in article)

    Oil Gas J. 67 #3:48 (1/20/69)

THE ROLE OF TURBOEXPANDERS IN LOW-TEMPERATURE PROCESSING IS GROWING. A survey covers the two basic low-temperature refrigeration cycles, condensing and noncondensing, in which the turboexpander is used, the thermodynamic aspects, with formulas for the exhaust temperature and the developed horsepower; the most important applications of the turboexpander, including the use of expansion turbines in air-separation plants, in the petrochemical field (especially ethylene plants), in the recovery of helium from natural gas, in the liquefaction of natural gas, in new processes for the tonnage production of liquid hydrogen, and in the recovery of propane and other hydrocarbons from natural gas; the mechanical design of cryogenic turboexpanders; and the control and safety devices. Photographs, diagrams, and graph.

    L. Atwood (Worthington Corp.) Oil Gas J. 67 #3:58-63 (1/20/69)

                    [16-1256]

STRUCTURE AND PROPERTIES OF MIXED LITHIUM-CALCIUM SOAP GREASES. A series of grease samples were prepared by compounding the oil MVP with 13% of a mixture of lithium and calcium stearates in mole ratios ranging from 19:1 to 1:1. The replacement of lithium stearate with equimolecular amounts of calcium stearate up to a mole ratio of 3:1 had no appreciable effect on the thickening action of the soaps and on the structure and rheological properties of greases. However, beyond the critical concentration of 35-40% calcium soap in the thickener, significant changes in structure were noted, along with deterioration of the rheological properties and the colloidal and mechanical stability. The addition of calcium soaps in lithium greases improves their water resistance and gradually reduces their dropping point. Table, graphs, photomicrographs, and 11 references. (in Russian)

    G. I. Svishevskaya, N. E. Shchipina et al. (Berdyansk Petrol. Lubric. Oil Works; All-Union Sci. Res. Inst. Petrol. & Gas Ind.) Khim. i Tekhnol. Topl. i Masel 13 #11:17-21 (1968)

                    [16-1300]

Fig. 7.5. Highlighting the first sentence. *Source. Abstracts of Refining Literature,* February 10, 1969. American Petroleum Institute. Reprinted with permission.

A $65 MILLION, 1000 METRIC TON/DAY ANHYDROUS
AMMONIA PLANT is being planned for Indonesia. Pertamin,
the Indonesian national oil and gas state enterprise, and
Universal Chemicals Ltd. (Nassau, Bahamas) have agreed
to build and operate the complex. It will include a single-
train ammonia plant in combination with units to convert
its ammonia into urea and nitrogen solution fertilizers.
The companies expect to produce urea for less than
$75/metric ton, compared with the present price
$85/metric ton for delivered urea. (no more info in article)
    Chem. Week 104 #4:44 (1/25/69)   [16-1391]

ALASKAN LNG TO JAPAN NEARS REALITY. Under
contracts signed by Phillips Petroleum Co. and Marathon
Oil Co. with Tokyo Electric Power Co. Inc. and Tokyo Gas
Co. Ltd. in Mar. 1967, 50 MMMCF/yr of natural gas will
be delivered to Japan over a 15 year period. The project
will require the design and construction of a
140 MMCF/day liquefaction plant and of deep-water docking
and loading facilities at Port Nikiski on Cook Inlet, the construction, in Swedish shipyards, of two 440,000 bbl LNG
tankers, and the design and construction of receiving and
regasification facilities in Japan. The first tanker will un-
load on schedule in July 1969. The communications problems, the present project status, some details on the
tanker construction and the receiving terminal, and some
unusual features involving investment costs and industry-
government relations are also discussed. Photographs.
    G. L. Farrar. Oil Gas J. 67 #3:28-30 (1/20/69)
                                                [16-1246]

EUROPEAN AUTOS OUTRUN FUEL QUALITY. According
to F. B. Fitch and R. H. Thena (Mobil Oil Co.) at a recent
SAE meeting (Detroit Jan. 1969), a recent substantial increase in the antiknock needs of European cars can be
attributed to changes in engine design to improve efficiency
and power. The study included the major car-building
countries of Australia, France, Italy, Japan, U.K., U.S.,
and West Germany, which together account for over 90%
of Free World passenger-car production. Outside the
U.S., the average vehicle octane number requirement at
the 50% satisfaction level has gained 6 to 10 points since
1958, whereas in the U.S., it has gained only 1 point in
Research octane number during the same period. The
present range of average premium-grade Research octane
numbers is 98-100, compared with 90-98 in 1957, but
premium gasolines in the U.K., Australia, France, and
West Germany satisfied less of the total car population in
1967 than they did in 1958.
    Oil Gas J. 67 #3:42 (1/20/69)   [16-1292]

                Fig. 7.5.   (Continued)

cialized audiences also call for some degree of orienting. For example, an abstractor writing for a microbiological abstracting service must consider the special needs of his audience. He can assume that most of the readers of his abstracts probably expect emphasis on the microbiological aspects of whatever is abstracted. And, more likely than not, they will expect these aspects to be covered in more depth than in a generalized abstracting service.

On the other hand, the abstractor who writes for one of the more general, large abstracting services in chemistry, biology, medicine, physics, or engineering can also orient his abstract, but he may do so in a more general way. A chemist who abstracts for a major generalized chemical service, such as *Chemical Abstracts,* would ordinarily look at an article in a physics journal primarily for new information of interest to chemists. But a physicist abstracting the same article for *Physics Abstracts,* for example, would look primarily for new information of interest to physicists.

Subject matter is closely related to the question of orienting. Abstracts can vary in construction and content, depending on subject matter. Different kinds of data are important in different disciplines.

For example, a chemist working in the field of organic synthesis would expect an abstract in his field to contain pertinent data on the compounds made, especially melting point, and on conditions and results such as temperature, pressure, yield, and percent conversion.

Similarly, workers in other fields of science and technology would expect abstracts in their fields to reflect the kinds of data considered important in that field to be reported. A biologist looking at an article also of interest to chemists would probably be much more interested in the biological effects of the compound studied than in the chemistry of the compound.

7.8. TREATMENT AND USE OF GRAPHIC MATERIAL

Should conclusions from tables, photographs, diagrams, and other graphic material be developed by the abstractor? We think the answer is "no." The author of the original document should have drawn his own conclusions from which the abstractor, as well as any other reader of the document, can work. But the abstractor can refer to the presence of graphic materials when these are significant. And, as previously mentioned, he may want to make use of graphic material both as a word- and work-saving measure in the construction of abstracts and for more effective presentation.

Abstractors should also take full advantage of graphics to enable them to read documents more quickly. Usually the graphic conveys its message

E3238 A VINYL LAMINATED NYLON BAG
is being used as an integral part of
a transport system which permits a van to
carry bulk materials one way and general
cargo on an immediate return trip. The
system was designed by Sani-Bulk Div.,
FLORIG EQUIPMENT CO. REVERE PLAS-
TICS fabricates the LINER from the laminate by a combination of electronic welding
and sewing of seams. A POLYESTER
THREAD is used where sewing is necessary, and the seam is sealed with a liquid
vinyl to cover needleholes. (more) *Mod.
Plast. 45, Sept. 1968 p132*

E3254 A RAYON MONOFILAMENT-REIN-
FORCED POLYESTER FILM TAPE
with tensile strength of 240 lb./in. has been
developed by BORDEN Inc. for heavy-duty
strapping and package reinforcing. The
tape is said to be resistant to shattering as
well as breaking. The tape is designated
MYSTIK 6481. (more) *Mod. Plast. 45, Sept.
1968 p246&248*

E3277 JAPAN GAS-CHEMICAL can get more
PARA-XYLENE from xylene fractions
by ISOMERIZING the less-valuable META-
XYLENE in the mixed stream. JGC is now
building a commercial plant in Japan that
should be onstream before year's-end. BAD-
GER CO. has been granted LICENSING rights
in the U.S. for the new extraction and separation process. *Chem. Week Sept. 14,1968
p79&80*

Fig. 7.6. Highlighting key words. *Source. Plastics Industry Notes,* September 23,
1968, and April 27, 1970, a publication of the Chemical Abstracts Service, published
by the American Chemical Society. Reprinted with permission of the Chemical
Abstracts Service of the American Chemical Society.

E1370. A new URETHANE FOAM INSULATION was invented by the Bilt-Foam Systems Division of TECHNOLOGICAL PRODUCTS, INC. J. Commer. April 10, 1970 p5

E1371. MONSANTO CO. has introduced a new grade of high-density POLYETHYLENE to serve as a high performance SHEET EXTRUSION RESIN. The new plastic is designated MPE 500. J. Commer. April 10, 1970 p5

E1372. Two new PLASTICIZERS have been introduced by the industrial chemicals division of GEIGY They are REOMOL L7-9 and REOMOL L9-11. Plast. Rubber Wkly April 10, 1970 p1

E1373. A machine that is used for disinfecting medical equipment and carries highly corrosive formaldehyde and ammonia vapors in the disinfecting process incorporates PVC PIPEWORK. The pipework is manufactured by DURAPIPE AND FITTINGS LTD. Plast. Rubber Wkly April 10, 1970 p1

E1374. PERMALI's PERMACLAD EPOXY GLASS COPPERCLAD LAMINATE for use in the printed circuitry industry, has been successfully tested to meet both British Standards and American Military Standard specifications. Plast. Rubber Wkly April 10, 1970 p1

E1379. For the fashion world of bees, soft pastel blue PLASTICS are in and the white painted wood is out. The man setting the fashion trend is scientist, Dr. G. A. Hobbs of the CANADA DEPARTMENT OF AGRICULTURE's research station, who is experimenting with POLYSTYRENE BEEHIVES. Plast. Rubber Wkly April 10, 1970 p10

E1380. An elastomeric WATERPROOFING MEMBRANE, BETASEAL 200, is being manufactured by the BFC Division of ESSEX CHEMICAL CORP. as a water barrier for exterior walls of foundations, cavity walls, under roof flashing, and between two-course concrete slab construction of roof decks. It is a two-part POLYSULFIDE MEMBRANE. Plast. Rubber Wkly April 10, 1970 p11

E1381. The MUDGUARDS of the BRAMBER SPRINGBOK TRAILER have been vacuum formed from black high density POLYETHYLENE SHEET extruded by TELCON PLASTICS LTD. The forming is carried out by HALLPLAS LTD. Plast. Rubber Wkly April 10, 1970 p11

E1382. A CRASH HELMUT that breaks new ground with a molded ABS OUTER SHELL and inner LINER of expanded POLYSTYRENE has been launched by THETFORD MOULDED PRODUCTS. It is being marketed under the trade name CENTURION X. Plast. Rubber Wkly April 10, 1970 p11

E1375. Concrete floors in the tannery of C. B. DANIELS LTD., in the UK, are protected with a TERCOL EPOXY PLASTICS system. _Plast. Rubber Wkly April 10, 1970 p4_

E1376. Significant cost savings have been claimed for a manufacturer of PNEUMATIC TUBE systems which has used GLASS REINFORCED PLASTICS instead of brass for some of its tight radius curves. The company is LAMSON ENGINEERING CO. LTD. _Plast. Rubber Wkly April 10, 1970 p4_

E1377. The cold flow and impact resistance and the molding properties of NYLON 6.6 have led an Austrian manufacturer to select this material for an ingeniously designed DOWEL CONNECTOR. The connector is employed for the rapid assembly of modular furniture units. _Plast. Rubber Wkly April 10, 1970 p9_

E1378. A RUBBER FACED TYPE that combines precision with flexibility has been announced by REJAFIX LTD., which claims that it has special application in the direct overprinting of cylindrical objects such as ampoules, vials, glass tubes and light bulbs. _Plast. Rubber Wkly April 10, 1970 p9_

E1383. A new technique for imparting a durable, abrasion resistant, nonstick coating to small components has been developed by K. AND F. TREATMENTS. It is based on the company's ARMOURCOTE process in which the substrate is first frame sprayed with stainless steel forming a spongy layer with excellent keying properties for the subsequent PTFE COATING. The new process, called MICROCOAT, uses IMPERIAL CHEMICAL INDUSTRIES LTD.'s FLUON PTFE. _Plast. Rubber Wkly April 10, 1970 p28_

E1384. The introduction of large diameter extruded POLYPROPYLENE TUBE with wall thicknesses up to 1 in. and pressure ratings up to 142 psi, means that material can now be used for the fabrication of corrosive fume and liquid handling plant where pressure limitations have previously excluded it. One of the first installations completed in this material is a PHOSGENE SCRUBBER manufactured by PLASTIC CONSTRUCTORS LTD. at the IMPERIAL CHEMICAL INDUSTRIES LTD. Rocksavage Works, Runcorn. _Plast. Rubber Wkly April 10, 1970 p28_

E1385. CROSS CUTLERY, the latest line in DISPOSABLES from CROSS PAPERWARE LTD. had been accepted by the COID for inclusion in Design Index. It is designed in white POLYSTYRENE. _Plast. Rubber Wkly April 10, 1970 p28_

Fig. 7.6. *(Continued)*

99

E1386. MARLEY TILE CO. has introduced SUPERF-
LEX, which is a VINYL TILE that Marley says
embodies all the important advantages of BS
3261. Plast. Rubber Wkly April 10, 1970 p28

E1387. PEPSI-COLA CO., in Las Vegas, is test-
marketing Pepsi-Cola in 10-oz. BOTTLES trade-
marked PLASTASTIC. The material has an ACRY-
LONITRILE base and was developed by the VISTRON
CORP., a subsidiary of STANDARD OIL CO. molding
is being done by VINYL MAID CO. Plast. World
April 1970 p4

E1388. In the BOEING 747, the 24 WING FUEL TANK
ACCESS PANELS, injection-molded of GLASS REIN-
FORCED NYLON 6/10, are believed to be the first
application of PLASTICS in the exterior body
structure of a plane. The material is LIQUID
NITROGEN PROCESSING CORP.'s THERMOCOMP QF-100-
10. Plast. World April 1970 p4

E1389. CADILLAC PLASTIC AND CHEMICAL CO. has
entered the ranks of basic raw-material producers. Cadillac will make OILON PV 80, an ACETAL
BASED RESIN compound with a lubricant and other
fillers. Plast. World April 1970 p8

E1390. DIAMOND SHAMROCK CHEMICAL CO. is developing markets for its new POLY(VINYL FLUORIDE)-
2 material, DALVOR. Plast. World April 1970 p8

E1391. A new cross-linkable high-density POLYE-
THYLENE RESIN for rotational molding has been
announced by PHILLIPS PETROLEUM CO. It is
called MARLEX CL-100. Plast. World April 1970
p9

E1396. A spokesman of BROOKHAVEN NATIONAL
LABORATORY announced a newly developed PLASTICS
IMPREGNATED CONCRETE. Plast. World April 1970
p15

E1397. Low-cost prefabricated HOUSING PANELS
have been developed by UNIVERSAL PAPERTECH CORP.
The core of the panels is PAPERBOARD called
UNIKRAFT. Panel interiors are coated with
FIBERGLASS MAT impregnated with POLYESTER. A
7-inch overlap of FIBERGLASS REINFORCED PLASTIC
extends from the edge of each panel. Plast.
World April 1970 p15

E1398. Articulated POLYETHYLENE PIPE made in a
continuous programmed sequence with straight
sections, corrugated intervals and end fittings
has been developed by Action Plastics Co., a
Div. of DART INDUSTRIES. Plast. World April
1970 p14

E1399. SHIP A SHORE CORP. has introduced a
23-ft COMBO CRUISER that accommodates a family
of six and can be towed by a standard automobile
on any especially built low-level trailer. Both
HULL and CABIN are vacuum-formed from a single
sheet of ROYALEX ABS supplied by UNIROYAL.
Plast. World April 1970 p15

E1400. LAWRY'S FOODS, INC. has developed a
sauce to ease taco preparation and a DYLITE FOAM
BOX that nests and protects 200 taco shells in
shipment. SINCLAIR-KOPPERS supplies the foam.
Plast. World April 1970 p16

E1392. AMERICAN CAN CO. is sealing beer and beverage cans by the MIRASEAM process, which uses a special lap seam and PARA BOND THERMOPLASTIC ADHESIVE. Plast. World April 1970 p11

E1393. GOODYEAR TIRE AND RUBBER CO. unveiled a new RUBBER and PLASTICS SAFETY GLASS replacement. Believed by Goodyear scientists to be stronger and lighter than previous glazing materials, the composite has rubber particles, invisible and offering no hindrance to light transmission, suspended throughout the PLASTICS RESIN. Plast. World April 1970 p12

E1394. Five students of PRATT INSTITUTE have designed a BOTTLE for salad oil. The container is blow-molded of modified PVC. Plast. world April 1970 p13

E1395. The elevator in the newly opened AUTOMATION HOUSE in New York has the first FRP CAB in the country. The cab was supplied by NATIONAL ELEVATOR CAB AND DOOR CO. The GLASSFIBER REINFORCED POLYESTER construction was selected over other materials because of its resistance to vandalism damage, light weight, ease of cleaning and ease of repair. Plast. World April 1970 p13

E1401. Twelve illuminated stained-glass-like BUILDINGS made of GLASSFIBER REINFORCED HETRON POLYESTER from HOOKER CHEMICAL form a three-story, 260-ft.-long VILLAGE OF LIGHTS display at CARSON PIRIE SCOTT AND CO. Plast. World April 1970 p16

E1402. A new two-way REFLECTIVE MARKER that signals a red warning to motorists traveling the wrong direction at night on divided highways or entrance/exit ramps has been introduced by the Stimsonite Div., AMERACE ESNA CORP. The raised marker consists of two REFLECTOR LENSES made of ROHM AND HAAS' PLEXIGLAS ACRYLIC set back-to-back in a cast-steel frame. Plast. World April 1970 p16

E1403 The TRACKSTER all-terrain vehicle manufactured by CUSHMAN MOTORS features a REINFORCED PLASTIC BODY with a FOAM-filled vacuum-formed ACRYLIC FLOTATION RING from AMERICAN CYANAMID CO. Plast. World April 1970 p17

E1404. LAMINATED PLASTICS SURFACINGS, used principally in wall, tabletop and countertop applications, have been fabricated into a set of MUSICAL STRINGED INSTRUMENTS. LEE STRACK has made a Hawaiian steel GUITAR, a rhythm guitar and a VIOLIN, and is currently working on an electronic BASS. LAMINATED PLASTICS SHEETS are supplied by CONSOWELD CORP. Plast. World April 1970 p29

Fig. 7.6. *(Continued)*

101

at a glance. The verbal explanation is often slower and more time-consuming.

7.9 THE ONE AND THE MANY

In his writing of abstracts, the abstractor should always remember that while many abstracts may stand alone, many others do not. For example, in literature surveys as described in Chapter 11, the grouping, arranging, writing, and indexing of many abstracts to form a coherent unit is often as important as the form and content of the individual abstract. The same comment applies, but to a lesser extent, to in-house patent and journal literature bulletins as described in Chapter 10.

chapter 8

Some Special Cases

8.1. WRITING MEETING-PAPER ABSTRACTS

Many meeting papers are never published. The quality of the abstract is especially important in such cases.

Because of the time factor, writing of meeting-paper abstracts may precede completion of all experimental work. The abstract may be shy of details—but the author might need to change these anyway later because of results from his latest experimental work or because of new ideas and insight. Many authors find that a certain amount of leeway can be most helpful by permitting them to present their latest and best thinking.

Some authors apparently believe that presentation of too much detail in their meeting-paper abstracts may discourage attendance at their oral presentations. The reasoning is that if there may be little of additional significance that a potential attendant could learn by being present in person there will not be the good-sized audience that most authors desire.

If he wants to attract an audience, the author of a meeting paper should also make his abstract *interesting*. If the abstract writer, who is almost always the author of the paper, is enthusiastic about the importance of his work, his abstract should reflect this. The author knows, of course, that if he overdoes it, his reputation will suffer. Similarly, the author owes it to his audience to present a paper that is consistent with the theme of his abstract. The bane of the meeting chairman is the author who delivers a paper unrelated to his original abstract.

Some sort of balance must be struck. The author of the meeting-paper abstract must provide useful, interesting, and thought-provoking information. But he must avoid falling into the trap of revealing too much, perhaps prematurely, and perhaps in an overly long abstract. In addition, the abstract should also be clear enough so that the paper can be properly placed in the program along with other papers on related aspects of the subject.

Some examples of meeting paper abstracts are shown in Fig. 8.1.

21. HELIUM, FROM RARE TO WELL DONE. C.W. Seibel, U.S. Bureau of Mines (retired), Amarillo, Texas 79102

Helium had been found in natural gas! The brief announcement carried in the minutes of the 33rd General Meeting of the American Chemical Society, December 29 - January 2, 1906, was destined to become the forerunner of the Federal Government's present multi-million dollar helium industry. The present paper reviews the development of this once rare gas from a chemical curiosity to its present role, that of an irreplaceable material in wide use by science and industry. Included is the story of its discovery in the "Wind Gas" of Dexter, Kansas, and other interesting highlights, such as who suggested its use for filling balloons when there was not a cubic foot available in the entire United States. The part the Bureau of Mines has played in initiating the production of helium from natural gas and in the continuing sponsorship of the project is told in detail. The first use of helium as a lifting gas for dirigibles is recounted, along with how important that use became in World War II. After the war, new uses for helium were developed so rapidly that additional means of producing it became necessary. Finally, the factors that pointed to the necessity for the present Helium Conservation Program are outlined, together with the story of how the Program was implemented and what it was designed to accomplish.

5.   METHODS OF PREPARING THERMALLY RESISTANT FIBERS.
Samuel C. Temin, Fabric Research Laboratories, Inc., Dedham, Mass. 02026

The methods used to prepare thermally resistant fibers from wholly aromatic or polyheterocyclic polymers are discussed. Because these polymers decompose before melting, only solution spinning processes have been utilized or reported. The discussion is limited, therefore, to a consideration of dry and wet spinning techniques and a brief description of the kind of equipment used. An elementary review of fiber formation by these techniques is given along with rule-of-thumb guidelines to assist those not skilled in the art to gain a better understanding and appreciation of the approaches and difficulties involved. Some of the experience reported in the literature with thermally stable fibers is used for illustration. Mention is made of the relative merits of dry and wet spinning as applied to aromatic polyamides, polyimides and other polyheterocyclic polymers. A brief description is given of equipment used for small-scale or screening examination of fiber formation when only limited quantities of polymer are available.

Fig. 8.1. Examples of meeting paper abstracts. *Source. Abstracts of Papers,* ACS National Meeting, Atlantic City, N. J., September 1968. Published by the American Chemical Society, Washington, D.C. Reprinted with permission.

81. METABOLIC ROLE OF MEDIUM CHAIN FATTY ACIDS. Sami A. Hashim, Dept. of Medicine, St. Luke's Hospital Center and Institute of Nutrition Sciences, Columbia University, New York, N.Y. 10025.
The fatty acid constituents of naturally occurring triglycerides are predominantly long chain. A small fraction(5-7%)is comprised of acids shorter than laurate. It has been possible to prepare medium chain triglycerides(MCT), composed mainly of octanoate and decanoate, for nutritional and metabolic studies in man and animals. Digestion, absorption and transport of MCT differ markedly from those of long chain triglycerides(LCT). Higher rates of pancreatic enzyme hydrolysis and intestinal absorption have been found for MCT than for LCT. An hydrolytic system for MCT has been identified in the mucosa of the small intestine, distinct from pancreatic lipase. After absorption, medium chain fatty acids (MCFA)are transported via portal vein bound to albumin. In contrast to the behavior of long chain acids, MCFA reaching the liver are almost quantitatively oxidized regardless of the nutritional state of the animal. However, extrahepatic tissues, if given the opportunity, as in portacaval shunting, physiologic hepatectomy, or parenteral infusion, are capable of oxidizing MCFA. MCT preparations are useful in management of chylous fistulas and fat-induced hyperlipemia. Also, a variety of malabsorptive disorders are amenable to treatment with MCT. These involve the gastrointestinal tract, the pancreas, the biliary system, or metabolic derangements associated with malabsorption. The usefulness of MCT is related to: high caloric density, low melting point, relatively small molecular size, a certain degree of finite solubility in water; low requirement for bile acids for micellar solubilization, efficient digestion and absorption, portal venous transport and extensive oxidation in the liver. Under normal conditions, MCFA are not stored in appreciable quantities in extrahepatic tissues.

139. APPLICATION OF INFRARED AND MASS SPECTROSCOPY TO THE IDENTIFICATION OF ORGANOPHOSPHORUS COMPOUNDS. August Curley and Robert Hawk, National Communicable Disease Center, Toxicology Laboratory, Pesticides Program, Atlanta, Ga. 30333

Recent work in this laboratory combined IR and mass spectroscopy to determine the structures of certain organophorphorus compounds. These compounds were either unknown technical products or residues. The use of these two powerful analytical tools together usually makes possible the complete assignment of structure of an unknown compound. To show the versatility and scope of the method, examples of structural determinations are given. There are some difficulties which can lead to erroneous conclusions in the interpretation of the IR spectra. The spectrum is often dependent on the method of sample preparation. Some types of organophosphorus compounds complex reversibly with certain solvents to yield adducts with spectra characteristically different from the original compounds. The IR spectral properties of several types of these heretofore reported adducts are described, and the probable electronics of the adduct formation are suggested. The mass spectra of the compounds are reported.

136. THE FLAVOR COMPONENTS OF ONION OIL, M. H. Brodnitz, C. L. Pollock, P. P. Vallon, and W.J. Evers, International Flavors & Fragrances, Inc., 1515 Highway #36, Union Beach, New Jersey 07735

The flavor components of onion oil were examined by combined gas liquid chromatography - mass spectrometry.

The various components of the oil were also separated by preparative-scale gas chromatography and examined individually by infrared, nuclear magnetic resonance and mass spectral techniques. Several sulfur-containing compounds not previously reported as occurring in onions were found to be the important components of their flavor.

Fig. 8.1. *(Continued)*

118. THE VOLATILE CONSTITUENTS OF JAPANESE HOP. <u>Yoko Naya,</u>
Yoshio Hirose, Munio Kotake, Institute of Food Chemistry, Osaka (Japan)

A volatile oil obtained from Japanese hop (Humulus lupulus L.) by steam
distillation and ether extraction of the steam distillate was fractionated into its
constituents by a combination of fractional distillation, column and preparative
gas chromatography.   Identification and characterization was carried out by
chemical and spectroscopic methods (mass, n.m.r., and infrared).   More
than one hundred compounds were identified, among which about one half were
newly found in hop oil.   Chemical classes mostly represented are oxygenated
terpenes, hydrocarbons, aliphatic esters, ketones, alcohols, lactones, and
aromatic and heterocyclic compounds.   In addition, the structural determination
of several new compounds was accomplished.   Details of the results will be
presented.

1. THE CHALLENGES IN MARKETING FOR AGRICULTURAL CHEMICALS. <u>J. F. Bourland</u>, American
Cyanamid Company, Berdan Avenue, Wayne, New Jersey, 07470

    With world food production due to rise dramatically, in what nations will the rise
be greatest? What will be the best locations for the plants to produce the agricultural
chemicals for these nations? In view of the far-reaching changes taking place in farms
and farming techniques, what marketing methods will best reach the large farm of
tomorrow without disrupting the network of distributors and dealers, which will be
needed for decades to come? What adjustments will be needed to meet the intensified
competition as patents on important agricultural chemicals expire? In the fertilizer
business, what adjustments will be necessary to bring the farmer the benefits of low
costs from the high capacities now building and to assure adequate return to the producer.

88. <u>DIELECTRIC EFFECTS IN ADSORPTION:</u> Carl Barlow, Jr.,
Texas Instruments Inc. Research
Labs., Dallas,Texas, 75200

How does adsorbed polarizable material alter the interaction between adsorbed ions?
The usual answer given to this problem is to introduce a dielectric constant, but
this simple approach is badly in error. This failure of the usual dielectric constant
concept has impact upon micro-potentials, virial coefficients, partition functions,
correlation functions and other functions relating to adsorbed systems. A new
method for treating properly the dielectric effects is introduced and applied to an
array of polarizations having no permanent dipole moment. The new approach
involves the introduction of a greatly generalized dielectric function, $\mathcal{E}$ (K, K')
which is essentially Green's function in reciprocal (Fourier-transformed) space.
This dielectric function provides a means for correctly treating polarization
effects, important local field corrections included.

Fig. 8.1. *(Continued)*

10. THE CHALLENGE IN MARKETING FOR DOMESTIC SALES IN PLASTICS. <u>B. W. Colaianni</u>, Chevron Chemical Co., Union, New Jersey.

The plastics industry is growing at the rate of 13 to 15 percent per year, compared with a rate of 4 to 5 percent for all other industry. This rate of growth for plastic industry is adding more volume and new products to meet this demand. Perhaps 50% of the products to be sold in the late 70's are unknown today, or are unavailable in large quantities. Such a dynamic industry--which is a major factor in our domestic economy--will demand a sales organization equally dynamic and innovated in its approach. It must be more adaptable, more flexible and be able to specialize its efforts in a manner not fully applied today. The new plastics sales organization of the 70's will move larger volumes of products, will sell more on a contract basis, and will direct its sales efforts based on centralized bulk terminals or off-plant manufacturing sites. Also, the more common use of distributors will come into play, the computer will help reduce costs and increase efficiency, and possibly it may even sell on a catalog type approach.

4. THE POST-WAR MODE OF DEVELOPMENT OF THE EUROPEAN PETROCHEMICAL INDUSTRY. <u>J. Moss</u>, Simon-Carves Chemical Engineerint Ltd., London, England, A. Doll-Steinberg, Simon-Litwin Ltd., London,England.

The development of the petrochemical industry in Europe after the Second World War, due to a number of factors, lagged considerably behind that of the USA, and due to the operation of some of the same factors, its present day shape differs in certain significant aspects from that of the USA.

These factors obviously include those stemming from the War itself, which while it effectively established the petrochemical industry in the USA, had in Europe a net retardant effect; in addition, there were and are the marked differences in demand -- both in its magnitude and in its pattern, the differences in raw materials availabilities and even the differences in the types of equipment available. No less important than these tangible differences, were the differences in technological attitudes and in attitudes towards investment, risk, obsolescence and innovation, and the Post-War political pressures in Europe.

The development of the European petrochemical industry, since the Second World War, up to the present time, will be described in terms of the above-mentioned factors and some forecasts of its future mode of development will be attempted in terms of the influences now bearing upon the industry.

12. MINIMIZING TRANSPORTATION COST. <u>O. Denton Hudson</u>, Armour Agricultural Chemical Company, Box 1685, Atlanta, Georgia 30301

Transportation charges comprise the major portion of distribution cost which most companies recognize as being one of the top three items of expense. Purchasing transportation needed at the most favorable cost, consistent with services required,is a complex technical undertaking and requires the talents of a knowledgeable traffic manager. Transportation should not be appraised on cost alone. It deserves a much broader view by management and a complete understanding of company operation and objectives by the traffic manager so that its relation to other functions is given proper consideration. Cost is not likely to be kept at a minimum on any other basis. There are many opportunities for the traffic manager to make valuable contributions to the company's economic well being in the areas of plant or distribution facility location, new product development and marketing, packaging and labeling, pricing structure, marketing programs, production scheduling, warehousing and distribution planning. There is also the ever present possibility of improving transportation cost thru favorable rate reduction or hold-down of general increases, as well as taking advantage of available alternatives in modes, type of service utilized and control of related cost.

Fig. 8.1. *(Continued)*

In some technical society meeting procedures, abstract length is controlled automatically by the requirement that all meeting-paper abstracts be submitted on a standard preprinted form supplied by the society. The form may have definite space limitations—often allowing for about 200 words maximum. The author must then be sure that his abstract is sufficiently compact to fit within the allotted space, which is often blocked out in rectangular form.

Abstracts are often reproduced photographically by the society sponsoring the meeting and distributed ahead of time to the membership. For this reason the author needs to be certain that the typing is neat. He also needs to be sure that the typewriter ribbon is fresh and that the keys are clean to ensure clear, bold reproduction. Use of a type size larger than is ordinarily used is sometimes appropriate.

The form used by the American Chemical Society for papers to be presented at its meetings illustrates the above points and is shown in Fig. 8.2.

8.2. WRITING AUTHOR ABSTRACTS

With the increasing tendency by abstracting services to use author abstracts, the author must usually write full abstracts, often to specifications. For example, see Fig. 8.3 for suggestions made by the American Chemical Society (11).

Requirements for author abstracts vary with the journal and field. Authors are well advised to consult the "instructions to authors" in recent issues of the specific journals in which they intend to publish.

8.3. WRITING HIGHLIGHT ABSTRACTS

In some technical and business publications (for example, *Paper Trade Journal, Chemical Engineering, Fortune*), abstracts (or really what more properly might be called "highlight abstracts") appear immediately after the table of contents or as part of the contents. Highlight abstracts are more "breezy," less formal, less complete, and are not intended for use by abstracting services. The intended audiences are often individuals who—with journal issue in hand—want to conserve their time by scanning highlight abstracts and reading only those articles of particular interest.

These abstracts may be prepared by either author or (more typically) journal editor; the editor will probably have more skill in this function. The highlight abstract should be prepared for easy reading, both in appearance (use of large type is recommended) and in wording. The reader

STANDARD ABSTRACT FORM FOR ACS MEETING PAPERS

Use attached form for an abstract which can be reproduced photographically, without retyping or retouching. DO NOT use black and white reproductions. Printed forms with nonreproducible blue lines are available from Division Secretaries and from the National Meetings Office at 1155 Sixteenth Street, N.W., Washington, D.C., 20036. The latter will also supply members, on request, with a free copy of Bulletin 8, "HANDBOOK FOR SPEAKERS."

Use the form to supply the following information required by the NATIONAL MEETINGS OFFICE

A NAME OF DIVISION

B TIME NEEDED TO PRESENT PAPER

C TITLE OF PAPER
Use Cap and Small letters for Benefit of Editors.

D AUTHORS' NAMES
First Names First.

E CURRENT ADDRESSES

F MEMBERSHIP STATUS
ACS & Division

G CLASSIFICATION & DEGREE

H WHERE WORK REPORTED WAS PERFORMED. *Omit if all authors are still there.*

A DIVISION OF

B Time Required: _____ Minutes

C TITLE OF PAPER

D AUTHORS
Underline name of speaker

E Complete Business Mail Address
List address only once if all authors at same address

F ACS Member?

G Division Member? American Chemist or Chemical Engineer?
If not, give classification such as biologist, physical etc. Ph.D*

H Work done at

I Plan ACS _____ nonACS _____ publication Where? _____

J Presentation requires projector: 2x2" (35 mm) _____ 3¼"x4" (standard) _____ Overhead _____ Vu-Graph _____ Film: 8 mm _____
16 mm _____ Sound _____ Other equipment (specify): _____

K ABSTRACT. 200 words or equivalent TITLE OF PAPER. Authors Names. Addresses with Zip Code. One-Line Space.
Abstract. Single-space typing Use full width of ruled area below.

Fig. 8.2. Standard abstract form for ACS meeting papers. *Source.* American Chemical Society. Reprinted with permission.

109

I. PUBLICATION INFORMATION

J. EQUIPMENT NEEDED FOR PRESENTATION OF PAPER

Follow the Guide for AUTHORS

K. HOW TO PREPARE A PHOTO-READY ABSTRACT
Format to use
Details to observe

L. WHERE TO MAIL FINISHED ABSTRACT
Also check for number of copies required.

"Be not careless in deeds, nor confused in words, nor rambling in thought."
—Marcus Aurelius
Meditations VIII, 51
A.D. 121-180

DO NOT USE →

DO NOT USE →

START THE ABSTRACT TITLE HERE USING CAPITAL LETTERS. Follow with Authors' Names, Business Addresses, Zip Codes. Underline Speakers' Name. Start third line and any subsequent lines in the heading, if needed, just inside the blue line at left.

Leave a space between heading and abstract proper. Indent as shown. Keep all lines as wide as possible without touching or going beyond the blue lines at either side. Short lines create extra pages and add to publication expense. Avoid them where possible. Keep the text in one paragraph. If literature citations are needed, insert them in parentheses and not as footnotes. Credits, if any, should be added at the end of the abstract, but not as a new paragraph. If structures or other forms of illustration are used, drawings should be part of the overall abstract, as shown below, and not submitted separately.

Use an electric typewriter with carbon ribbon if possible, and a type size to give about 88 characters (letters) per 7½ inch line. Before submitting your abstract, check format, nomenclature, and spelling. Make sure that erasures do not show. Abstracts will not be retyped, but reproduced photographically at two-thirds the original size, minus the blue guidelines which are nonreproducible. If the standard form is not available when you need it, use plain white paper. Do not draw guidelines. Set your typewriter for a 7½ inch line and use the format shown here. Please mail the abstract unfolded.

DO NOT USE →

L. MAIL ABSTRACT TO PERSON NAMED IN ACS DIVISIONAL DEADLINES PUBLISHED (JUNE & DEC.) IN C&EN

Time Required: _____ Minutes

DIVISION OF

TITLE OF PAPER

		ACS Member?	**Division Member?**	**American Chemist or Chemical Engineer?**
AUTHORS	**Complete Business Mail Address**			*If not, give classification such as biologist, physicist, etc. Ph.D?*
Underline name of speaker	*List address only once if all authors at same address*			

Work done at

Plan ACS _____ nonACS _____ publication. Where? _____ No _____ Uncertain _____

Presentation requires projector: 2x2″ (35 mm) _____ 3¼ x4″ (standard) _____ Overhead _____ Vu-Graph _____ Film: 8 mm _____

16 mm _____ Sound _____ Other equipment (specify) _____

ABSTRACT. 200 words or equivalent. TITLE OF PAPER, Authors' Names, Addresses with Zip Code, One-Line Space, Abstract. Single-space typing. Use *full width* of ruled area below.

Fig. 8.2. (*Continued*)

111

DO NOT
USE →

DO NOT
USE →

MAIL ABSTRACT TO PERSON NAMED IN ACS DIVISIONAL DEADLINES PUBLISHED (JUNE & DEC.) IN C&EN

Fig. 8.2. (*Continued*)

112

Abstract—Every article must be accompanied by an informative abstract that summarizes the principal findings of the work reported in the paper. Although usually read first, the abstract should be written last to ensure that it reflects accurately the content of the paper.

Through a cooperative effort between the primary publications of the American Chemical Society and the Chemical Abstracts Service, a drastic reduction in the time lag between journal publication and abstract publication in *Chemical Abstracts* is taking place. This is being accomplished through the direct use by *Chemical Abstracts* of the abstracts submitted with the original papers. In a majority of the cases, the abstract is being processed for publication in *Chemical Abstracts* at the same time that the original article manuscript is in press.

For this effort to achieve maximum success, it is essential that the author be aware of the new importance taken on by his abstract. The ideal abstract will state briefly the problem, or the purpose of the research when that information is not adequately contained in the title, indicate the theoretical or experimental plan used, accurately summarize the principal findings, and point out major conclusions. The author should keep in mind the purpose of the abstract, which is to allow the reader to determine what kind of information is in a given paper and to point out key features for use in indexing and eventual retrieval. It is never intended that the abstract substitute for the original article, but it must contain sufficient information to allow a reader to ascertain his interest. The abstract should provide adequate data for the generation of index entries concerning the kind of information present and key compounds.

The abstract should be concise. Only in unusual cases should it contain more than 200 words. The nomenclature used should be meaningful; that is, standard systematic nomenclature should be used where specificity and complexity require, or "trivial" nomenclature where this will adequately and unambiguously define a well-established compound. References to numbered figures, tables, or structures presented in the body of the paper may be made in the abstract because these may readily be incorporated when the abstract is used in *Chemical Abstracts* (see Figures 1 and 2).

Chemical Abstracts indexing in the areas of synthetic and theoretical organic and inorganic chemistry is done from the original paper. For such papers the general content of the abstract should be as stated above, but the large number of compounds frequently encountered in papers in these areas precludes inclusion of all of them in the abstract. They are however included in the *Chemical Abstracts* index.

In general it is of utmost importance that abstracts be rich in indexable information to guide interested readers to the original paper.

The preceding examples illustrate useful, informative abstracts. The reader close to these topics knows at a glance whether the entire paper is of interest to him.

Fig. 8.3.   Instructions to authors on writing abstracts for their journal papers. *Source. Handbook for Authors of Papers in the Journals of the American Chemical Society,* Washington, D.C., 1967, pp. 20–22. Copyright 1967, American Chemical Society Publications. Reprinted with permission of copyright owner.

Hydrolysis of Phostonates[1,2]

Anatol Eberhard and F. H. Westheimer

*Contribution from the James Bryant Conant Laboratory
of Harvard University, Cambridge, Massachusetts.*
Received September 29, 1964

Abstract: A five-membered cyclic ester of a phosphonic acid, lithium propylphostonate (**1**), and a six-membered analog, lithium butylphostonate (**2**), have been synthesized. The rates of hydrolysis of these compounds, relative to that of sodium ethyl ethylphosphonate (sometimes compared directly, sometimes extrapolated to 75°) in acid are 5×10^4:3:1, and in alkali are 6×10^5:24:1. Tracer methods with ^{18}O show that the phostonates are cleaved at the P–O bond, whereas the hydrolysis of the open-chain phosphonate occurs with about half P–O and half C–O fission. The relative rates of hydrolysis at phosphorus are then slightly more favorable to the phostonates than the figures shown above. The previously established extraordinary reactivity of cyclic five-membered esters of phosphoric acid is thus paralleled by that of the cyclic phosphonates.

The extraordinarily large rates of hydrolysis of five-membered cyclic esters of phosphoric acid,[3] as compared to those of the six-membered cyclic esters,[4] or of the corresponding open-chain compounds stimulated an investigation of the properties of the corresponding phosphonates. The dilithium salt of 3-bromopropyl-phosphonic acid was cyclized to the phostonate, by

internal displacement. Similar reaction led to the

FIGURE 1.—Example of an abstract as published in the *Journal of the American Chemical Society* and, to the right, as published in *Chemical Abstracts*. Note the handling of references to structures.

Hydrolysis of phostonates. Anatol Eberhard and F. H. Westheimer (Harvard Univ.). *J. Am. Chem. Soc.* **87**(2), 253–60(1965)(Eng). A 5-membered cyclic ester of a phosphonic acid, Li propylphostonate (I), and a 6-membered analog, Li butylphostonate (II), have been synthesized. The rates of hydrolysis of these compds., relative to that of Na Et ethylphosphonate (sometimes compared directly, sometimes extrapolated to 75°) in acid are 5×10^4:3:1, and in alkali are 6×10^5:24:1. Tracer methods with ^{18}O show that the phostonates are cleaved at the P–O bond, whereas the hydrolysis of the open-chain phosphonate occurs with about half P–O and half C–O fission. The relative rates of hydrolysis at P are then slightly more favorable to the phostonates than the figures shown above. The previously established extraordinary reactivity of cyclic 5-membered esters of phosphoric acid is thus paralleled by that of the cyclic phosphonates. RCJC

Fig. 8.3. *(Continued)*

Reaction of Aryl Ketones with Cyclopentadienyl Sodium.
Syntheses of Fulvenylmethanols

R. J. MOHRBACHER, V. PARAGAMIAN, E. L. CARSON, B. M. PUMA, C. R. RASMUSSEN,
J. A. MESCHINO, AND G. I. POOS

Department of Chemical Research, McNeil Laboratories, Inc., Fort Washington, Pennsylvania

Received January 4, 1966

The reaction of 2-benzoylpyridine with cyclopentadienyl sodium in alcohol can be directed to give the expected 6-phenyl-6-(2-pyridyl)fulvene (3) as its dimer in 88% yield or the novel α-phenyl-α-[6-phenyl-6-(2-pyridyl)-2-fulvenyl]-2-pyridinemethanol (4) in 86% yield by varying the conditions. The reaction conditions which favor formation of 3 or 4 are discussed in terms of a mechanism for their formation. A variety of diaryl and alkyl aryl ketones, in which the aryl groups were phenyl, substituted phenyl, 2-, 3-, or 4-pyridyl, thienyl, or quinolyl, were allowed to react with cyclopentadienyl sodium. It was found that strongly electronegative aryl groups are required for conversion of diaryl ketones to 2-fulvenylmethanols. Aryl 2- (or 4-) pyridyl and di-2- (or 4-) pyridyl ketones form 2-fulvenylmethanols readily. Most diphenyl ketones do not form 2-fulvenylmethanols readily and alkyl pyridyl ketones give only trace amounts of fulvenylmethanols.

As part of our effort to synthesize 6,6-diarylfulvenes which are intermediates to bridged hydroisoindolines, 2-benzoylpyridine (1) was treated with cyclopentadienyl sodium (2) to give the anticipated 6-phenyl-6-

grated for one proton and (b) the higher melting isomer of 4 possesses the cis configuration.[6]

Early workers[7] had considered the possibility of ring-substituted fulvenes, such as 9, arising from reaction of

CHART I

rapid dimerization to 7,[10] or an 86% yield of 2-fulvenylmethanol 4 (condition A). The product compo-

In terms of the mechanism, the results of var the re   conditions (Table I) suggest that low

FIGURE 2.—Example of a concise abstract as published in *The Journal of Organic Chemistry* and, to the right, as published in *Chemical Abstracts*.*

Fig. 8.3.   (Continued)

* For corresponding abstract, see top of p. 116.

Reaction of aryl ketones with cyclopentadienyl sodium. Syntheses of fulvenylmethanols. R. J. Mohrbacher, V. Paragamian, E. L. Carson, B. M. Puma, C. R. Rasmussen, J. A. Meschino, and G. I. Poos (Dept. of Chem. Res., McNeil Labs., Inc., Fort Washington, Pa.). *J. Org. Chem.* **31**(7), 2149–59(1966) (Eng). The reaction of 2-benzoylpyridine with cyclopentadienylsodium in alc. can be directed to give the expected 6-phenyl-6-(2-pyridyl)fulvene (I) as its dimer in 88% yield and the novel α phenyl-α-[6-phenyl-6-(2-pyridyl)-2-fulvenyl]-2-pyridinemethanol (II) in 86% yield by varying the conditions. The reaction conditions which favor formation of I or II are discussed in terms of

a mechanism for their formation. A variety of diaryl and alkyl aryl ketones, in which the aryl groups were Ph, substituted phenyl, 2-, 3-, or 4-pyridyl, thienyl, or quinolyl, were allowed to react with cyclopentadienylsodium. Strongly electroneg. aryl groups are required for conversion of diaryl ketones to 2-fulvenylmethanols. Aryl 2- (or 4-) pyridyl and di-2- (or 4-) pyridyl ketones form 2-fulvenylmethanols readily. Most diphenyl ketones do not form 2-fulvenylmethanols readily and alkyl pyridyl ketones give only trace amts. of fulvenylmethanols.

RCKF

Fig. 8.3. *(Continued)*

expects such abstracts to be factual, informative, and if he is to go on to read the full text of the original article, interesting. Figures 8.4 and 8.5 illustrate the points made.

Another good example of this practice is the succinct daily "What's News" column of business and other news which appears on page 1 of *The Wall Street Journal.*

8.4. WRITING ABSTRACTS FOR IN-HOUSE REPORTS

In most organizations the preferred practice is to ask each author to prepare his own abstracts for reports he writes describing the results of his research. Report authors are usually asked to follow a *Standard Practice,* which outlines specifications for all parts of the report including the abstract. For example, a typical set of instructions might read as follows:

The abstract is intended to convey succinctly the objectives, scope, and general findings of the report to those who may not read the full report. It must not exceed one typewritten page and must be on a separate page. For purposes of identification the following information must appear at the top of the page: full title of report, author(s), time period covered by report, report number, and project number. The abstract

The Special Case of Specialty Steels 129

A small group of specialty-steel companies—Latrobe, Carpenter, Allegheny Ludlum, Cyclops, and Crucible—are bringing a luster to their names that the great tonnage producers must surely envy. Last year Carpenter earned 17.3 percent on invested capital, more than three times U.S. Steel's paltry showing. These companies haven't by any means avoided all the ills that beset the steel industry, notably rising labor costs, but they do point the way to dodging some of them. They have geared their plants to produce stainless for atomic reactors, superalloys for jet engines, and tool steels for machine-shop drills, and in the process commanded prices as high as $7 per *pound*. Their attention to high technology and custom work makes them much less susceptible to import assaults, and their small size and diverse product lines make them less vulnerable to government pricing pressure. Equally important, the specialty-steel makers are selling to the economy's growth industries such as aerospace and electronics.

U.S. Science Enters a Not-So-Golden Era 144

In the wake of federal budgetary cutbacks, the U.S. scientific community suddenly faces an uncertain future after years of steeply rising appropriations. Many Congressmen have turned not only parsimonious, but skeptical, and one has even declared that basic research may be too expensive a hobby in an age that calls for quick solutions to immediate problems. The squeeze is being felt most severely at the universities. For lack of funds, hundreds of graduate students in the physical sciences are being turned away and others, already on the campus, may lose their stipends. If the squeeze continues, the country faces a severe shortage of scientists in the years ahead, and the effects will be particularly acute in the fields of chemistry and physics. Dr. Jerome B. Wiesner, provost at M.I.T., warns that unless current trends are reversed, the U.S. will become "a very sick country" technologically.

Unfortunately, American scientists neither have a strategy for allocating today's limited appropriations nor a clear notion of scientific priorities. A new approach must begin with an appraisal of our scientific goals and a plan to provide for their continuing financial support. One current proposal: peg research and development expenditures to a fixed percentage of the gross national product, such as 3 percent.

Fig. 8.4. Highlight abstracts (for tables of contents of journals). *Source. Fortune,* November 1968. Copyright 1968, *Fortune,* New York. Reprinted by permission of copyright owner.

Proposals for using short take-off (STOL) and vertical take-off (VTOL) airplanes, which got almost nowhere in the past, are now receiving serious attention as air-traffic delays squeeze the profitability out of conventional short-haul flying. Eastern Air Lines already is flying a STOL plane on its shuttle runs on an experimental basis. A host of air-taxi and commuter airline entrepreneurs are putting small foreign-built STOL planes to work. And both Los Angeles and New York are proposing to build multimillion-dollar STOLports close to their central business districts in the next few years.

Two fundamental problems remain, and solutions are several years away: designing a commercially acceptable plane that can take off and land at slow speeds and cruise at high ones; and providing navigation equipment that can make more sky room available for the large volume of flights that is expected.

Fig. 8.4. *(Continued)*

Highlights Of This Issue

DEVELOPMENTS IN ENERGY CONVERSION—86

New technology is sparking progress in energy-conversion developments that may lead to commercial applications. Of particular interest are the marriage of the internal combustion engine with the electric car, and other hybrid systems; new organic fluids for use in a Rankine cycle; and new structural and fuel systems.

RISING TIDE OF DESALTING PLANTS—90

As older desalting methods move into large-scale units, newer techniques still in pilot stage continue their progress toward a competitive position. Result: an almost-twentyfold increase in desalting capacity by 1980.

ENGINEERING AWARD WINNERS NAMED—96

Four distinguished practitioners of chemical engineering have won CE's first biennial award, established to honor the individual engineer and his personal excellence in his chosen profession.

CATALYST HELPS XYLENE ISOMERIZATION—138

In the novel Isoforming process, use of a catalyst that contains one or more nonnoble metals does much to improve the production of paraxylene. The catalyst costs less than its noble-metal counterparts, and cuts down on unwanted side reactions.

THERMOPHYSICAL PROPERTIES OF LIQUIDS—152

Introducing a new series in which many of the most widely used methods for estimating thermophysical properties of liquids are statistically evaluated with respect to accuracy and reliability. Recommendations emerge as to what methods are best for each class of liquid.

FOCUS ON EMPLOYEE BENEFITS—155

A survey of 64 CPI companies reveals policies for such benefits as pension plans, vacations, additional-compensation programs, hospitalization and life insurance.

WORKING WITH THE PROTOTYPE PLANT—163

Prototype plants are built as a stage between the pilot plant and the full-scale plant. This article contains hints on how to handle such a project.

THE BASIS FOR BODE PLOTS—177

Frequency-response tests are frequently used to establish the dynamics of interacting components in many physical systems, or the dynamics of the system itself. To analyze the data from such tests, appropriate Bode diagrams must be established—this article shows how.

DIATRIBE OF A TECHNICAL EDITOR—184

Engineers have a well-deserved reputation for being poor writers. Many companies, and virtually all publications, provide editors to help the engineer communicate properly. This article humorously describes the varieties of engineers who resent this help.

MEASURING TEMPERATURE: NO CONTACT—188

Process temperatures that cannot be measured or monitored with conventional instruments, because the target cannot be touched, can be measured and monitored with infrared radiation instruments.

USING SUPPLY AND DEMAND CURVES—198

The relationship between the supply and demand curves for a product determines how its profitability varies with production volume. The author presents a technique for investigating the profitability of a venture when the demand curve can be defined.

Fig. 8.5. Highlight abstracts (for tables of contents of journals). *Source. Chemical Engineering,* October 7, 1968, p. 3. Copyright 1968, by McGraw-Hill, Inc., New York, N. Y., 10036; reprinted by special permission from *Chemical Engineering.*

should not contain unidentified trade names, code names, or unusual terms or abbreviations.

In a well-administered procedure for report writing, the appropriate level of management reviews and approves the full text, including the abstract.

Those individuals immediately involved in a project almost always want and get the full report. Some organizations, however, prefer to circulate abstracts only (instead of the full text) to those whose interest is less direct or not immediately obvious. This can save substantial reading and filing time, as well as paper, printing costs, and space. In such cases, the full text of the report should be readily available in a technical report center for easy consultation.

Report authors should be made aware that their report abstracts can be a principal means for making the quality and importance of their work known to management. Although some executives will get the abstract only, even those who get the full report may not ordinarily have the time or interest to go beyond the abstract.

The report author can, with his abstract, affect the executive in several important ways. He can encourage the executive to look more fully into the report. Or he can cause an executive to dismiss work as not significant. Also, if the abstract is not clearly written, the executive will be confused and will probably not be inclined favorably toward the report.

All of this means that the author of in-house reports must write abstracts so as to encourage further interest, especially if the results are indeed noteworthy. But the abstractor must not lead the executive to expect what is not there or something that does not clearly stem from the full text of the report. Most executives appreciate clear and interesting abstracts, and they appreciate equally objectivity and accuracy, both in the report itself and in the abstract.

Some reports may also include a summary, which should not be confused with the abstract. A typical set of instructions for a summary in a *Standard Practice* for writing in-house reports might read as follows:

This part of the report contains substantially more information than is found in the abstract. It should include: a statement of the broad objectives; a summary of previous work and reports; objectives of this particular report; and principal findings. Other general introductory and historical information can be included here. Information should be given on the reported work's specific contribution toward achieving project goals.

Unlike the abstract, the summary is not ordinarily distributed separately from the entire report.

Like the full reports, both abstracts and summaries of in-house reports are proprietary to the organization. Circulation and access are limited accordingly.

8.5. ABSTRACTING PATENTS

Patents make up an important body of literature since they constitute the basis of much of the growth of science and technology, both in the United States and abroad. It is beyond the scope of this book to go into detail about patents although some basic information is appropriate. This information is intended for abstractors and should not be regarded as a legal interpretation. The reader interested in more information about patents can consult references (9) and (10) and should also contact the United States Patent Office for further detail.

Patents are unique sources of technical information in that they are also legal documents. Many patents are difficult to read and abstract precisely because they are legal documents, which are usually written in legal phraseology by the attorneys who act as agents for the inventors.

In the United States the granting of a patent (under the laws in effect in 1970) indicates that the inventor has the right to exclude others from using his invention for a period of 17 years after the patent is issued by the United States Patent Office.

For those abstractors who are not familiar with patents, a brief outline of some of the parts of a United States patent may be helpful. These include:

1.   The *drawings*.

2.   The *abstract*.

3.   The so-called *specification*, which constitutes the main body of the patent in terms of size.

4.   The *claims*, which always appear at the end of the patent and delineate the scope of the monopoly.

If the abstractor reads the abstract and the first few paragraphs of the specification, he will be able to get some idea of the general purpose and nature of the invention. He should look for sentences beginning with such phrases as: "The purpose of this invention is," in order to identify precisely what is invented.

The specification also usually contains in the first few paragraphs some indication of the uniqueness (that which is new) and of the advantages of the particular invention over "prior art"—that which is already known.

In many patents the specifications include "examples," which are so labeled. One of these examples can be used by the abstractor with some degree of assurance that this will be useful to his clients since "examples" usually give specific experimental details.

Claims are often quite broad (generic) and highly legalistic in phrasing. Claims often express succinctly the essence of the patent and are of special interest to management, patent attorneys, and scientists in industrial organizations.

The challenge faced by the abstractor of patents is that of understanding the legalistic terminology and phrasing, and of extracting information that can be used in an abstract.

The abstractor may want to use a combination of claims and examples. The claims characterize the scope of the invention, and the examples supply the detail. Here is such a composite:

> A novel process for the mfg. of product XYZ is claimed. For example, X, Y, and Z are reacted at 100°C at atmospheric pressure in the presence of catalyst A. XYZ is obtained in 80 percent yield.

This is the kind of abstract that is probably a happy compromise for many patents.

Figure 8.6 shows examples of abstracts prepared by Derwent Publications, Ltd., a leading patent-abstracting service, which now covers patents in almost all technologies. The abstracts illustrate not only abstracting of patents but also some of the points made elsewhere in this book:

1. Often somewhat telegraphic writing which conserves space.
2. Specific, meaningful titles (especially important with patents).
3. Use of graphics to save space and convey data quickly.
4. Format convenient for clipping and filing into personal files.

All applicants for United States patents are now required to submit abstracts along with their applications. The abstracts of issued (accepted) United States patents are published each week in the *Official Gazette of the United States Patent Office*. Guidelines for writing abstracts for this purpose are shown in Fig. 8.7.

Figure 8.8 contains examples of abstracts from the *Official Gazette*. The use of diagrams intended to clarify the abstract is noteworthy. The small numbers identify the various parts of the diagram and are explained in the full text (specification) of the original patent.

Figure 8.9 explains and illustrates a proposed format for the first page of United States patents. This figure indicates most of the major parts of a patent (except the full specification) and also shows clearly the difference between a claim and an abstract. (See p. 144).

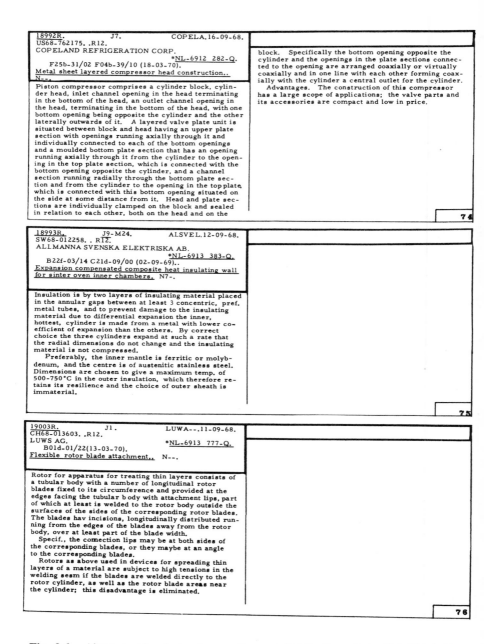

Fig. 8.6. Abstracts of patents. *Source.* Derwent Publications, Ltd., 128 Theobalds
Rd., London, WC 1, England. Reprinted with permission of copyright owner.

19004R.   H6-J7.   AIR LIQ.13-09-68.
FR68-166209. . R12.
L'AIR LIQUID SA.   *NL-6913 785-Q.
  F25j-03/08 (17-03-70)..
Separation of gas mixtures esp carbon dioxide and
nitrogen from natural gas. N6B.

Gas mixtures are separated by washing the gas mixture
with a wash fraction in which the unwanted components
are soluble, this fraction being produced in liquid state
from the separation phase flowing downwards from the
washing, and containing practically no unwanted components.

The gas mixture is cooled by heat exchange with
fractions produced in the separation phase, before it
is used in the washing operation. The gas mixture may
also be cooled by heat exchange with the wash fraction
which is enriched in unwanted components, or by heat
exchange with the purified gas mixture.

Specifically
The purified gas mixture may be at least partly condensed by heat exchange with the wash fraction contg.
the unwanted components, or by heat exchange with an
external cooling source, and/or by evaporation of the
was fraction, or by the liquid fraction produced in the
separation phase. The wash fraction is pref. com-

pressed before washing.

Uses/Advantages
The process is especially useful for treating natural
gas, for the removal of carbon dioxide, nitrogen and
water vapour. The process is capable of efficiently
removing nitrogen and carbon dioxide from natural gas
by washing with liquid methane. The ratio of carbon
dioxide and nitrogen to methane in the natural gas may
be low or high, thus indicating the flexibility of the
process. (19004R).

77

19011R.   H6-J7.   AIRPOR.16-09-68.
GB68-044009. . R12.
AIR PRODUCTS LTD.   *NL-6913 880-Q.
  F25j-03/02 (18-03-70)..
Liquefaction of gas mixtures esp natural gas,. N6B.

Liquefaction of a gas mixture is carried out by means
of a cooling fluid consisting of a mixture of gases whose
components are also components of the gas mixture to
be liquefied, but not necessarily in the same proportions
while at least one foreign gas, which is not present in
the gas mixture to be liquefied, is added to the cooling
fluid in order to increase the yield.

Pref. the gas mixture to be liquefied in natural
gas, and the added foreign gas is ethylene. The proportions of butane and propane in the cooling fluid gas
mixture are so adjusted that the cooling and heating
curves of the system are correctly thermodynamically
related. The cooling fluid gas mixture is cooled, after
compression, by means of a closed recirculation system
in which one of the mixture components is used as
coolant. The mixture component is pref. butane.

Advantages
The method is especially suitable for liquefying
natural gas, using the components of the natural gas,
i.e. propane and butane, as the cooling fluid. The use

of a foreign gas, such as ethylene, enables the yield of
the process to be increased. The use of a closed recirculation system increases the yield, and also provides independent control of the temp. of the cooling
fluid gas mixture. (19011R).

78

19012R.   E19-I4.   MITCHE.14-09-68.
JA68-065948. (+18-02-69/JA69-012083). R12.
MITSUBISHI CHEMICAL INDUSTRIES LTD.
                                *NL-6913 892-Q.
  C07c-45/12 (17-03-70)..
Recovery of rhodium catalyst from oxo-reaction prod.
uct. N5-.

Reaction of olefins with CO and H_2 in presence of a
rhodium catalyst results in production of rhodium
compound. Action of water or steam at 100°-250°C
and pressure of 1-30 atmos., in presence of inert gas
(N_2 or H_2) and a suitable catalyst carrier releases
rhodium metal to the carrier. Method allows continuous reaction and regeneration of the catalyst. Fresh
rhodium salt solution added as needed to keep catalyst
concentration constant.

Prior Art
During the reaction rhodium metal is lost as rhodium carbonyl. Due to cost of rhodium, recovery of the
catalyst efficiently is essential for use on a large
scale.

Concentration
of rhodium in reaction product is 0.01 - 1 g/litre,
thus requiring 5% by weight carrier to reaction product

for recovery of the rhodium metal. Rhodium separated
from the carrier by burning to oxide or recovery as a
salt.

In an example
140 cc. hexene-1 and 40 mg. of 1% -rhodium-active
charcoal catalyst were treated at 110°C with CO and
H_2 at total pressure 240 atmos. for 2 hours. Regeneration of the catalyst was then effected by adding 10 cc.
water and heating to 100°C for 2 hours, under an atmosphere of N_2. 97% of the rhodium was recovered and
the catalyst was then ready for re-use. (19012R).

79

Fig. 8.6. *(Continued)*

19615R. E23-E36-G6.	EASTMA.23-09-68.	G6-A, G6-H7A, G6-H7B, G6-H 4

19615R. E23-E36-G6.
US68-761841. . R12.
EASTMAN KODAK CO.   *BE-0739 220-Q.
(ogilivra-).
(02-03-70)..
Sensitising photographic emulsions with dyes absorbed
on silica. M3B.

Sensitizing dye is adsorbed on finely-divided silica and
added to silver halide emulsion.
   The dye is prevented from migrating except to
neighbouring Ag halide crystals, and use of organic
solvents for manipulating dyes is eliminated.
Preferably
   Silica particles are 0.001 to 10 microns average
diameter.
   Dyes are cyanines and merocyanines, particularly:
1,1'-diethyl-2,2'-cyanine, the hydroxide of anhydro-5,5'-
6,6'-tetrachloro-1,1'-diethyl-3, 3'-di(3-sulphobutyl)
benzimidazolocarbocyanine, or 3(dimethylaminopropyl)-
-5-[(3-methyl-2-thiazolidinylidene)-ethylidene/-2-
-thio-2,4-oxazolidinedione.

Process
   Dye (0.001 to 0.200 g per gram-atom Ag) is dissolved

G6-A, G6-H7A, G6-H7B, G6-H   **4**

in an organic solvent and mixed with silica. the solvent
is removed and the dyed silica is added to liquid emulsion
at 40-60°C. Any Ag halide dispersed in any usual medium coated on any usual backing is suitable.
   The emulsions can be sensitized with a sulphur compound or with noble-metal compounds. Additives e.g.
anti-fogging agents can be used. The emulsions can be
tanned and are usable in colour processes. (19615R)

8 6

19616R. E37-G5.
US68-761852. . R12.
EASTMAN KODAK CO.   *BE-0739 221-Q.
Yackel EC. (yackeleec).
(02-03-70)..
Nucleation of aluminium lithographic plates.. M2B.

Grained Al plate is treated with basic aqueous solution
and aqueous Ag salt solution, to obtain receptive support
for a diffusion-transferred image.
   Plate treatment other than graining (e.g. anodising)
is unnecessary; no colloidal binder is needed, and
water-soluble Ag salts can be used; complete operation, in two steps or one, completed in 60 sec.
Preferred
   Single-stage treatment using basic solution of Ag
nitrate, acetate ot lactate containing ammonium, sodium
or potassium hydroxide.
More specifically
   Any water-soluble silver salt may be used; other
cations can replace Ag as long as they are more electro-
positive than Al.
   Two-stage treatment can use alkali or alkaline earth
hydroxides, oxides or anhydrides in alkaline aqueous
solution, specifically decinormal KOH, NH₄OH, NaOH;
neither concentration nor pH (range 10-14) are critical.

G5-A1, G6-A4   **2**

Single-stage treatment best uses silver salt dissolved
in ammonium hydroxide and then added to NaOH solution.
   Process ensures that the quantity of Ag nuclei deposited is of little importance; solutions containing 0.2
to 0.0005 moles/litre Ag can be used.
   Plate can be grained by etching or by mechanical
brushing or milling.
Examples
   (1) Al plate of thickness 0.127 mm, grained by
brushing, is immersed in 0.5% aqueous NaOH for 30
seconds at ambient temperature, then rinsed and
immersed in 0.1 N Ag acetate for 30 seconds at ambient
temperature, rinsed and dried. Plate is then covered
with emulsion, exposed, developed, and further treated
(British Pat. 934,691).
   Alternatively, Ag solution is (2) 0.1 N lactate, (3)
0.1 N nitrate; (4) KOH used instead of NaOH.
   (5) to 0.1 N AgNO₃ is added ammonia solution to
precipitate and redissolve silver hydroxide, and 0.1 N
NaOH is added equivalent to amount of Ag hydroxide.
Al plate is immersed for 1 minute, rinsed and dried.
(19616R)

8 7

19622R. E16-G6.
CH68-014256. . R12.
CIBA SA.   *BE-739 245-Q.
(23-03-70)...
Use of perfluoroaliphatic wetting agents in photographical
compositions, M3B.

Agents are non-ionic or anionic componds of formula
R$_f$-Z,(where R$_f$ is a perfluoroalkyl group; and Z an optionally substituted carboxylic or sulphonic residue) dissolved in gelatine solutions more particularly solutions,
cpds. of formula R$_f$-SO₂-N -R₁-CO-O-X (where R₁ is
             R₂
alkylene, pref. -(CH₂)₁₋₆-; R₂ is H, or alkyl, pref. methyl
or ethyl; X is H or alkali metal; R$_f$ is pref. F-(CF₂)₃-9-
or H - (CF₂)₃-9. The preferred compound is F-(CF₂)-
SO₂-N ── CH₂-COOK.
    C₂H₅

Most of the compounds are known, and commercially
designated as e.g. "FC" and Zonyl S". Use of the compounds is as wetting agent in photographical compositions, e.g. films, optionally in conjunction with an azo
pigment, in solutions of gelatin; processes of obtaining such compositions; and the compositions themselves
are claimed.

G6-E4, G6-H18   **2**

Advantages
Gives rise to homogeneous films, without streaks
etc., and hence improved final images.
Compounds
   (a) anionic:- R$_f$-COOX; R$_f$-SO₂-OX;

R$_f$-CO-N-R₁-COOX and esp. R$_f$-SO₂-N-R₁-COOX;
   R₂   R₂

(b) non-ionic:- R$_f$-SO₂-N-R₃; R$_f$-CO-N-R₃; R$_f$-CO-OR₄;
            R₂   R₂

R$_f$-O-CO-R₃ and R$_f$-O-CO-R$_f$. (R₃ = H or alkyl; R₄ =
alkyl).
Example
   On a cellulose triacetate support was spread a photographic emulsion containing 25 g. of Silver (as bromide
+ 2.5% iodide) and 40 g gelatin and 0.05-3.0 g.
"FC128" ("probably (I) above") per kg. of emulsion. All
layers were faultless and uniform. Sensitometry showed
no variation in sensitivity, gradation or opacity. (19622R)

8 8

Fig. 8.6. *(Continued)*

124

Photographic material contains an organic photoconduct-
or, a chromogen and a substance which produces free
radicals when irradiated. Pref. 1-30% wt. of free-rad-
ical former is used, based on photoconductor. The
photoconductor is at least one vinylcarbazole, pref. a
halovinylcarbazole or one of their copolymers.

Preferably
The vinylcarbazole is a (co)polymer or a vinylcarbazole and at least one of aromatic amine derivs. di-
phenylmethanes, triphenylmethanes, or is one or more
of these cpds. alone.
The photosensitive material incorporates an organic
chromogen constituent chosen from secondary or tertiary arylamines, carbazoles, indoles, 1,3,4-oxadiazoles,
1,3,4-triazoles, imidazoles, pyrazoles, oxazoles substituted by an aminophenyl radical, 1,3-di-
phenyltetrahydro-imidazoles, phenadines, acridines,

G6-C4, G6-F3, G6-F6, G6-H1   (CANO)   4

acylhydrazone derivatives of quinoxalines, acrydenaz-
ines containing a double substitution amino radical in
the N position, pyridines, quinolines, ethylene derivs.
oxazole derivs. derivs. of triazoles or imidazoles and
organic photo-chromic compounds. (19666R).

89

Photosensitive material produced by reacting organic
photoconductive substance, substance giving free-rad-
icals when irradiated, and dye base compound.
Gives sensitivity equal to or better than Se or ZnO,
with transparency, flexibility, lightness, film-forming
properties and selectivity to polarity of charge characteristic of organic photoconductors.

Preferably
The substance giving free radicals is 1 to 30% by wt.
of the organic photoconductor.
The photoconductor is at least 1 vinylcarbazole, pref. a
(co)polymer of a vinylcarbazole opt. halogenated; and/or
an aromatic amine deriv. or diphenylmethane or tri-
phenylmethane deriv.
The basic dye is a leuco- or carbinol base, styryl
base, merocyanine or leuco-dihydroanthracene.

Free radical source
A cpd. RCX_3 (where R = H, Cl, Br, I, opt. substituted

G6-C4, G6-F3, G6-F6, G6-H1, G6-H7 (CANO)   5

alkyl, opt. substituted aryl, or aroyl; X (similar or
different) = Cl, Br, I, halogenated sulphone. or halogenated sulphoxide or a halogenated organic compound, e.g.
a halogenated sulphoxide.
Particularly, RCX_3 is carbon tetrabromide or iodo-
form; the halogenated free radical source is tribromo-
sulphone; the aromatic amine photoconductor is 4,4'-
-bis-dimethylaminobenzophenone.
The free radical source may be irradiated first or
after compounding with other materials. The irradiation
effect may be increased by heating. (19667R)

90

Maintaining a minute deposit of fluorinated hydrocarbon
polymer on the photoconductive surface to prolong its
life.
By making the cleaning brush on roller of fluoropolymer fibres, a separate device for depositing fluoropolymer to prevent build-up of a retained layer of
developing powder is rendered unnecessary.

Preferably
The fibres are arranged in tufts on a cylindrical
supporting core. The fluoropolymer deposit is less
than 1 micron thick.

Advantage
In normal use, the developing powder tends to build
up on the drum over a period of time, giving a grey
background to the prints; this build-up, possibly due to
frictional fusion by the removing brush, is prevented by

G6-G8   1

a molecular layer of fluoropolymer, as previously
known; using fluorocarbon polymer fibres in the removing brush is novel.
A roller covered with tufted fabric of fluoropolymer
can also be used.

Example
Developer is pigmented powder of butyl methacrylate
-polystyrene copolymer carried on sand. Brush turns
at 1250 r.p.m. and contacts drum over 0.3 cm. No
developer film present on drum after 12,000 reproductions; a white film of fluorocarbon polymer appeared
after about 200 cycles but had no adverse effect on
reproduction. Substituting a brush of vinylchloride-
acrylonitrile copolymer fibres, a film of developer
appeared on the drum after 2,000 reproductions and
after 3,000 necessitiated a drum change. (19673R)

91

Fig. 8.6. *(Continued)*

125

19147R.　　　　J4.　　　　　　/LISTH.30-04-68.
OE68-004207.. R12.
LIST H.　　　　　　　　　　*FR-2007　328-Q.
　G01n-27/00 A61b-05/00 (02-01-70).
Device for PH determination of blood sample.　.F--.

Devices comprises three-way stopcock with plug which
has electrolyte chamber.　In this is end of reference
electrode.　Chamber has admission and flow openings.
These are closed in first plug position join admission
and electrolyte flow channel in intermediate position
and in second position, entry opening·to electrolyte
chamber is closed and flow opening joins lateral opening to capilary measuring tube.　This has advantage
of simple rapid manipulation and precise measurement
of pH of blood sample.

91

19156R.　　　　J4.　　　　　　INDFIL.05-04-68.
US68-719154. . R12.
INDUSTRIAL FILTER AND PUMP MFG CO.
　　　　　　　　　　*FR-2007　389-Q.
　Schmidt-H Zievers-JF. (schmidzie).
　B01j-04/00 B01j-01/00 (09-01-70).
Liquid delivery tube,.　F--.

Liquid contains material to be treated and passes under
pressure down delivery pipe with closed end.　Pipe
has several holes and fine mesh sleeve fits over it.
Along outer wall of tube, extended piece fits over
holes and between sleeve and tube, keeping sleeve apart
from hole in outer wall of tube.
　Advantages are uniform distribution of liquid and
material without blocking holes in delivery tube.
Applications include filtering resins and pulverised
matter.

92

Fig. 8.6.　*(Continued)*

126

19160R.   E 36-J1.   USATOM.16-04-68.
US68-721676.. R12.
US ATOMIC ENERGY COMMISSION.
                  *FR-2007 459-Q.
  Babcock-DF. (babcockdf).
  B01d-59/00 (09-01-70)...
Improved method of producing heavy water by isotope
exchange at two temps. F--.

Two gas absorption contacting towers are used at
different temperatures in order to concentrate an isotope
of an element by its exchange between a liquid phase and
a gaseous phase containing the element. The liquid
phase enters the first column with a natural concentration of isotope which becomes enriched by preferential
exchange of isotope with the gaseous phase in this first
column.

On passing to the second column the concentration of
isotope in the liquid phase is reduced by reversed
preferential exchange at a different temperature to a
concentration below its initial value. The liquid phase
is taken off as waste effluent. The gaseous phase
recycles in counter-flow with the liquid through the
columns, a proportion being tapped off to obtain concentrated isotope.

The improvement is to introduce a further supply of
the liquid phase in the lower part of the second contacting column.

The isotope deuterium is concentrated by exchange
between water and gaseous H_2S in the above process.

Detail

The plant comprises a hot column and a cold column
Water is supplied to the top of the cold column by a pipe
with a natural concentration of D:H, that is about
1:7000. In the cold column, this concentration of isotope progressively increases by preferential transfer
from the H_2S in counter-flow.

Waste effluent drawn off at the base of the hot tower
contains a lower concentration of deuterium than the
initial supply water. Isotope-enriched H_2S recycles
from the hot column, through the cold column and back
again in a closed circuit apart from a tap-off point for
humidified, hot enriched gas and a return point for
recooled gas.

The water should be introduced at a level at which
isotope concentration is just reverting to a level corresponding to that of the freshly introduced water.

The additional water should be preheated to about the

9 3

same temperature as the liquid in the hot column at the
point of addition.

Pref. the additional water represents between 10%
and 200% of the quantity of water introduced at the top
of the cold column. (19160R)

Fig. 8.6. (Continued)

19161R.   E36-J1, K8.   USATOM.16-04-68.
US68-721675. . R12.
US ATOMIC ENERGY COMMISSION.
                          *FR-2007 461-Q.
  Babcock-DF. (babcockdf).
  B01d-59/00 (09-01-70)..
Isotope enrichment by gas/solvent exchange..   F6C.

In the process, two gas absorption contacting columns are used at different temperatures, one hot and one cold, in order to concentrate an isotope of an element by its exchange between a liquid phase and a gaseous phase containing the element. The liquid phase enters the first column with a natural concentration of isotope which becomes enriched by preferential exchange of isotope with the gaseous phase at the temperature prevailing in the column.

At the different temperature in the second column, reverse isotope exchange takes place to such an extent that the liquid phase reaches an isotope concentration level below that of the initial supply liquid, and is drawn off as waste. The gaseous phase recycles in counterflow with the liquid through the columns, a proportion being tapped off to obtain concentrated isotope.

The essential improvement is to split the liquid phase discharge from the first column and to pass the total gaseous phase from the second column through part only of the liquid phase, in a contacting chamber.

Output and/or concentration of the isotope, usually deuterium, is thereby substantially increased.

Preferred process

The process is esp. applied to concentration of deuterium by exchange between water and H_2S as follows:-

$$H_2S + HDO \underset{cold}{\overset{heat}{\rightleftharpoons}} H_2O + HDS$$

The current of water is separated into two, one of which is contacted with all the H_2S enriched at the temp. of the cold tower, at least part of the current extracted as product or feed for a further stage. The second current is fed to the second tower. The contact of the first separate current is carried out in $< 1/3$ of the tower, pref. the lower $1/3$. (19161R).

94

19180R.   E36-J4.   KNAPSA.26-04-68.
DT68-767322. . R12.
KNAPSACK AG.   °FR-2007 536-Q.
  C01b-17/00 C01b-25/00 (09-01-70)..
Phosphorus pentasulphide prodn by reaction of the
elements above 300 degrees centigrade.   F6A.

The reaction is carried out in more than one reaction zone linked to a separate receiver which is in turn linked to a cooling vessel. The intermediate receiver and the connecting pipes leading to the other vessels are heated. Pref. two reaction zones are used and the product from each is collected in a single receiver from which it passes to the cooler. The rates of addn. of starting material and removal of product are such that the period spent in the receiver by the reaction prods. is 0.5-12 h, pref. 1-2 h. Pref. the receiver and the pipes connecting it to the reactor and cooler are kept at 340-380°C, pref. 350°C by electrical heating.

Advantages

Preparation of P_2S_5 by the conventional process tends to give rise to blockages in the connecting pipework, esp. at bends and valves, when short interruptions in the flow of the molten product occur due to breaks in production. Overheating these sensitive sites does not appreciably reduce the incidence of blockage but tends to increase the iron content of the P_2S_5 which can rise to 10%. Use of two reactors feeding into a common receiver eliminates the risk of flow interruptions and also reduces the time of contact of the molten product with the receiver to 0.5-12 hrs., normally 1-2 hrs., before discharge to a suitable grinding unit, with a consequent reduction in final iron content.

Example

Comparison of processes : In the conventional process P is reacted with S at 380-400°C and discharged from the receiver after a contact time of 3-4 hrs. In the improved process larger quantities of P and S can be reacted in two reactors discharging into the common receiver, the contact time with which only being 1-2 hrs. The conventional process gives a product contg. 14.1 ppm iron but the product from the improved process only contains 7 ppm. The improved process also gives a more homogeneous product, in terms of P content (28.0%), than the former. (19180R)

95

Fig. 8.6.   (Continued)

128

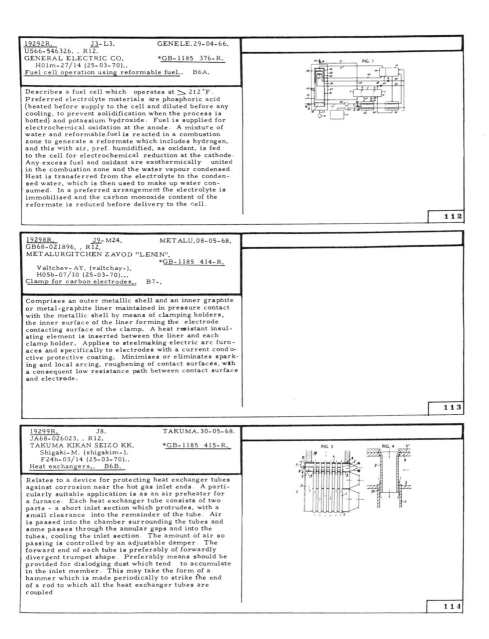

19292R.　　J3-L3.　　GENELE.29-04-66.
US66-546326. . R12.
GENERAL ELECTRIC CO.　　*GB-1185 376-R.
　H01m-27/14 (25-03-70)..
Fuel cell operation using reformable fuel.　B6A.

Describes a fuel cell which operates at $\geq 212°F$.
Preferred electrolyte materials are phosphoric acid
(heated before supply to the cell and diluted before any
cooling, to prevent solidification when the process is
hotted) and potassium hydroxide. Fuel is supplied for
electrochemical oxidation at the anode. A mixture of
water and reformable fuel is reacted in a combustion
zone to generate a reformate which includes hydrogen,
and this with air, pref. humidified, as oxidant, is fed
to the cell for electrochemical reduction at the cathode.
Any excess fuel and oxidant are exothermically united
in the combustion zone and the water vapour condensed.
Heat is transferred from the electrolyte to the condensed water, which is then used to make up water consumed. In a preferred arrangement the electrolyte is
immobilised and the carbon monoxide content of the
reformate is reduced before delivery to the cell.

112

19298R.　　J9-M24.　　METALU.08-05-68.
GB68-021896. . R12.
METALURGITCHEN ZAVOD "LENIN".
　　　　　　　*GB-1185 414-R.
　Valtchev-AY. (valtchay-).
　H05b-07/10 (25-03-70)...
Clamp for carbon electrodes.. 　B7-.

Comprises an outer metallic shell and an inner graphite
or metal-graphite liner maintained in pressure contact
with the metallic shell by means of clamping holders,
the inner surface of the liner forming the electrode
contacting surface of the clamp. A heat resistant insul-
ating element is inserted between the liner and each
clamp holder. Applies to steelmaking electric arc furnaces and specifically to electrodes with a current condu-
ctive protective coating. Minimises or eliminates sparking and local arcing, roughening of contact surfaces, with
a consequent low resistance path between contact surface
and electrode.

113

19299R.　　J8.　　TAKUMA.30-05-68.
JA68-026023. . R12.
TAKUMA KIKAN SEIZO KK.　　*GB-1185 415-R.
　Shigaki-M. (shigakim-).
　F24h-03/14 (25-03-70)..
Heat exchangers.. 　B6B.

Relates to a device for protecting heat exchanger tubes
against corrosion near the hot gas inlet ends. A particularly suitable application is as an air preheater for
a furnace. Each heat exchanger tube consists of two
parts - a short inlet section which protrudes, with a
small clearance into the remainder of the tube. Air
is passed into the chamber surrounding the tubes and
some passes through the annular gaps and into the
tubes, cooling the inlet section. The amount of air so
passing is controlled by an adjustable damper. The
forward end of each tube is preferably of forwardly
divergent trumpet shape. Preferably means should be
provided for dislodging dust which tend to accumulate
in the inlet member. This may take the form of a
hammer which is made periodically to strike the end
of a rod to which all the heat exchanger tubes are
coupled

114

Fig. 8.6.　(Continued)

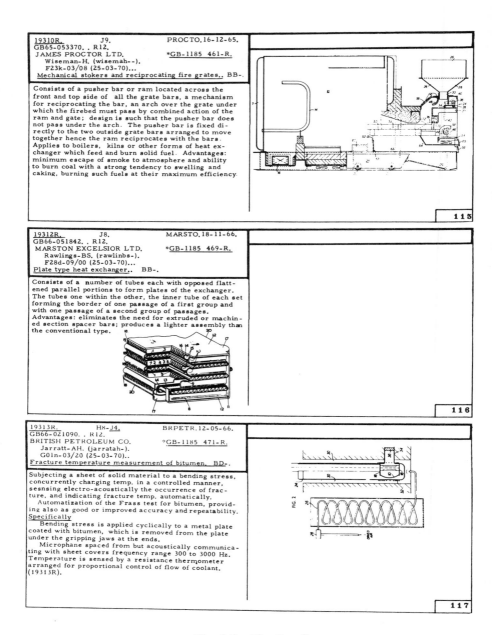

19310R. J9.	PROCTO. 16-12-65.

19310R.   **J9.**   PROCTO. 16-12-65.
GB65-053370. . R12.
JAMES PROCTOR LTD.   *GB-1185 461-R.
   Wiseman-H. (wisemah--).
   F23k-03/08 (25-03-70)...
Mechanical stokers and reciprocating fire grates.. BB-.

Consists of a pusher bar or ram located across the
front and top side of all the grate bars, a mechanism
for reciprocating the bar, an arch over the grate under
which the firebed must pass by combined action of the
ram and gate; design is such that the pusher bar does
not pass under the arch. The pusher bar is fixed directly to the two outside grate bars arranged to move
together hence the ram reciprocates with the bars.
Applies to boilers, kilns or other forms of heat exchanger which feed and burn solid fuel. Advantages:
minimum escape of smoke to atmosphere and ability
to burn coal with a strong tendency to swelling and
caking, burning such fuels at their maximum efficiency.

115

19312R.   **J8.**   MARSTO. 18-11-66.
GB66-051842. . R12.
MARSTON EXCELSIOR LTD.   *GB-1185 469-R.
   Rawlings-BS. (rawlinbs-).
   F28d-09/00 (25-03-70)...
Plate type heat exchanger.. BB-.

Consists of a number of tubes each with opposed flatt-
ened parallel portions to form plates of the exchanger.
The tubes one within the other, the inner tube of each set
forming the border of one passage of a first group and
with one passage of a second group of passages.
Advantages: eliminates the need for extruded or machined section spacer bars; produces a lighter assembly than
the conventional type.

116

19313R.   **H8-J4,**   BRPETR. 12-05-66.
GB66-021090. . R12.
BRITISH PETROLEUM CO.   *GB-1185 471-R.
   Jarratt-AH. (jarratah-).
   G01n-03/20 (25-03-70)..
Fracture temperature measurement of bitumen. BD-.

Subjecting a sheet of solid material to a bending stress,
concurrently changing temp. in a controlled manner,
sesnsing electro-acoustically the occurrence of fracture, and indicating fracture temp. automatically.
   Automatization of the Frass test for bitumen, providing also as good or improved accuracy and repeatability.
Specifically
   Bending stress is applied cyclically to a metal plate
coated with bitumen, which is removed from the plate
under the gripping jaws at the ends.
   Microphane spaced from but acoustically communicating with sheet covers frequency range 300 to 3000 Hz.
Temperature is sensed by a resistance thermometer
arranged for proportional control of flow of coolant.
(19313R).

117

Fig. 8.6. *(Continued)*

130

E1: GENERAL ORGANIC

E11: ORGANOPHOSPHORUS; ORGANOSILICON

21943R.   C1-E11.   MONCHE.01-12-62.
CH62-014134. .R14.
MONSANTO CHEMICAL COMPANY.
                            *US-3504 025-S.
    Maier-L. (maierl---).
    C07f-09/42 (31-03-70). (260-543)..
Aryl thiophosphorus compounds prepd in high yield from
thiophosphoryl chloride using friedel-crafts catalysts..

Compounds of formula: $Ar_mR_nP(S)X_{3-m-n}$ are prepd.
in high yield with minimal by-product formation by
reacting at a temp. at which hydrogen halide is given
off, and in the presence of a Friedel-Crafts catalyst,
a thiophosphorus halide of formula: $R_nP(S)X_{3-n}$ with
aromatic compd. ArH. (Ar = opt. aromatic radical;
pref. free of non-benzenoid unsatn; R = organic
radical; pref. free of non-benzenoid unsatn. and both
radicals Ar and R may contain substituents inert to
Friedel-Crafts catalysts; X = halogen, pref. Cl or Br;
m = 1-3; n = 0-2). It is essential that there is used
at least an equimolar amount of catalyst with respect
to halide to be reacted and at least an equimolar
amount of the ArH based on the halogen atoms to be
replaced. (26.11.63 as 326227) (MONS)

22035R.   E11-F6-H7-L1.   MIDSIL.23-05-66.
GB66-022921. . R14.
MIDLAND SILICONES LIMITED.   *US-3504 007-S.
    Cooper-BE Owen-WJ. (cooperowe).
    C07f-07/10 (31-03-70)..
Amino silanes from alkali metal amides and haloalkyl-
organosilanes..

Cmpds. (I) $R_aSi(ZNXX')_{4-a}$ (in which R is a mono-
valent hydrocarbon, halohydrocarbon or hydrocarby-
loxy radical; a is 1,2 or 3; X is a divalent saturated
aliphatic or cycloaliphatic radical; X is H, alkyl or
aryl and X' is alkyl or aryl) are prepared by reacting
a compd. (II) R_aSi ZY $_{4-a}$ (in which Y is halo)
with a compd. (III) XX'NCl (in which Cl is an alkali
metal atom) optionally in a solvent e.g. diethyl ether,
suitably at the reflux temperature using a stoichiometric excess of (III).
    Cmpds. (I) are useful in forming water repellent
coatings on fibres or as sizes for glass fibre, and
in improving oxidative stability or organic cmpds. e.g.
hydrocarbon. (17.5.67 as 644,427).

22039R.   D25-E11.   PROGAM.23-08-66.
US66-574470. . R14.
PROCTER AND GAMBLE COMPANY.
                           *US-3504 024-S.
    Diehl-F L Drew-HF Laughlin-RG. (diedrelau).
    C07f-09/54 (31-03-70). (260-526).
Phosphonio carboxylate detergent compositions effect-
ive in cool water..

New cpds:-
$$R^1 - \overset{\overset{R^2}{|}}{\underset{\underset{R^3}{|}}{P^+}} - (CH_2)_n - CO_2^- \quad\quad (I)$$

(where R^1 = 10-18C alkyl, R^2 and R^3 = 1-4C (hydroxy)
alkyl and n = 3-5), esp. 4-(P,P-dimethyl, P-dodecyl

phosphonio)butyrate (II), are effective detergents in
compositions with a water-sol. inorg. alkaline
builder salt (IV) in cool water (at 40-95°F, pref. 60-
90°F). Prepn. is by reacting a suitable tertiary
phosphene with a ω-chloro (or bromo) carboxylate
and saponifying, eg (II) was prepd. by refluxing
dimethyldodecyl phosphine with a mixt. of 4-chloro-
butyrate and MeOH for 5 days then saponifying the
ester with KOH in i-PrOH. (23.8.66 as 574470).

22048R.   E11.   MOBOIL.11-09-64.
US64-395933. (+06-10-67/US67-673309). R14.
MOBIL OIL CORPORATION.   *US-3504 055-S.
    Andress Jr-HJ Capowski-J. (andrescap).
    C07f-09/08 C07c-101 (31-03-70)..
Neutral primary amine salts of tripolyphos
phoric acid and phosphoric acid alkyl esters as hydro
carbon fuel stabilizers..

Useful stabilisers for liquid hydrocarbon fuels of
formula:

$$\left[R_1 - \overset{\overset{}{|}}{\underset{\underset{R_2}{|}}{N}} - R_3 \right]_n X$$

(where R = H or alkyl group of 8-18 C atoms., X = (a)
a phosphoric acid with 2-8 P atoms/mol. or
(b) a phosphoric acid alkyl ester with 2-8 P atoms/mol
and containing from 1-22 C atoms/alkyl group; n = 1-5)
    The cpd. is pref. a neutral primary alkyl amine salt
of tripoly phosphoric acid the primary alkyl amine
being a mixture of branched chain amines having from
about 8-18 C atoms and primary amine group directly
attached to tert. C atom. (6.10.67 as 673,309)
(E16).

22133R.   A18-E11-G2.   FMC---.07-04-67.
US67-629070. . R14.
FMC CORPORATION.   *US-3503 912-S.
    Lynch-JA. (lynchca--).
    C07g-01/10 (31-03-70). (260-028 260-027 260-029).
Levelling agent for emulsion floor polish...

Tris (2-hexoxyethyl) phosphate is used as levelling
agent (instead of tris (2-butoxyethyl) phosphate) in
polymer emulsion floor polishes containing a major
portion water-insoluble polymer, a waxy material
and a minor portion of an ammonia-soluble resin. 1/4 -
2% levelling agent is used, preferably 0.3-1.2%. The
polish flows out to smooth, glossy, uniform films of
excellent appearance and high gloss.(7.4.67 as 629,070)

22164R.   A17-E11.   KEUESS.21-12-67.
US67-692261.. R14.
KEUFFEL AND ESSER CO.   *US-3504 056-S.
    Lecher-HZ Braus-H Woltermann-JR.(bralecwol).
    C07f-09/08 C08f-45/58 (31-03-70). (260-945)..
Preparation of tri(dialkylaminophenyl thioalkylene)
phosphites and use as olefin polymer stabilisers..
Claim 1 is for phosphites

Fig. 8.6. *(Continued)*

$$\left[R_2N-\hspace{-4pt}\langle\hspace{-4pt}\bigcirc\hspace{-4pt}\rangle\hspace{-4pt}-S-(CH_2)_n \right]_3 PO_3$$

where n is an integer, 2-6, and R is a C1-4 alkyl, e.g.
(a) is tri(4-dimethylaminophenyl thioethylene) phosphite
where R = CH₃, and n = 2.
   These compounds are useful in compositions with
carbon black and/or hindered phenols in reducing heat
deteriorations of olefin polymers. (21.12.67 as 692261)
(KEUF)

22205R.   E11-F6.   STEVEN.02-03-67
US67-620109.. R14.
STEVENS AND CO INC JP.   *FR-1582 819-Q.
   Tesoro-GC. (tesorogc-).
   D06m B27k (01-09-69).
Fireproofing for cellulosics containing non-ionic
phosphorus and organic nitrogen.

In fireproofing cellulosics with non-ionic organic P
components, improvement is to use on substrate less
P than required for fireproofing and also organic N
compound in sufficient quantity to render the substrate
fireproof.
   Synergistic effect of N and P compounds gives
improved mechanical properties of treated textile
at lower cost. (26.2.68 as 141198)

22246R.   A15-E11.   ALBWIL.12-12-66.
GB66-055604. . R14.
ALBRIGHT AND WILSON (MFG) LIMITED.
                            *FR-1583 183-Q.
   Hoye-PAT Champman-AC Frank-G. (chafrahoy).
   C08f (24-10-69)..
Group 8 metal salts of alkyl thiophosphoric acids as
U V stabilisers for PVC..

Salts have the general formula $\left[(RY')(RY'')PXY'''\right]_2M$,
where R is /are 3-12C pref. 3-6C (cyclo) alkyl group(s)
X is S or O, or is absent; at least one of X, Y', Y'', Y'''
is S, the rest being O; M is Group 8 metal.
   0.01-0.5 pref. 0.1-0.5, wt.% of such compounds,
pref. with thermal stabilisers and anti-oxidants,
pref. 2,6-di-tert. - butyl-4-methyl phenol, give
slightly coloured blends with polymers containing
≥25 wt.% of vinyl chloride. (11.12.67 as 131,772)
(ALBR).

22458R.   A36-E11-G2.   BADANI.27-09-68.
DT68-793514.. R14.
BADISCHE ANILIN AND SODA FABRIK AG.
                            *NL-6914 494-Q.
   C07f-09/40 C07f-09/44 (01-04-70).
N,n - dialkoxymethylcarbamyl phosphoric acid dialkyl
esters. N2A.

Compounds with formula (I) prepared by reaction of
compounds with formulas (II a, b, c) with formaldehyde
in presence of base, followed by reaction with alcohols
of formula R₃OH in presence of acid.

$$R_1O\diagdown \overset{\displaystyle O}{\underset{\displaystyle |}{P}} \diagup -(CH)_n-CON\diagup^{CH_2OR_3}_{\diagdown CH_2OR_3} \qquad (I)$$
$$R_1O\diagup\qquad\quad R_2$$

where R₃ = alkyl, R₂ = H, alkyl, -OH or -Hal, n is 0-3.
R₁ is alkyl, or a heterocyclic 5 or 6-membered ring
with the P and O atoms.

$$R_1O\diagdown\overset{\displaystyle O}{\underset{\displaystyle |}{P}}\diagup-(CH)_n-CONH\quad-\quad H\qquad(a)$$
$$R_1O\diagup\qquad R_2$$
$$\qquad\qquad\qquad\qquad\qquad\qquad-CH_2OH\quad(b)$$
$$\qquad\qquad\qquad\qquad\qquad\qquad-CH_2OR_3\quad(c)$$

Compounds are used as additives to lacquers and syn-
thetics to increase flame resistant properties. A suitable flame resistant lacquer for metal surfaces prepared
from 90 parts by weight aminoplastic resin and 10 parts
by weight compound (I), treated with a hardening cmpd
and kept for a few mins. at 150-180°C. (24.9.69 as
6914494).

22466R.   A44-E11.   DYNNOB.28-09-68.
DT68-792651. . R14.
DYNAMIT NOBEL AG.   *NL-6914 639-Q.
   C01b-33/24 (01-04-70)..
Purification of chlorosilanes..   N6A.

(I) by (1) adding 0.01-5.0 mol.% (esp. 0.02-2.0 mol.%)
of a complexing agent comprising an N-heterocyclic
cpd. (II) with 1 or more nuclei and an S atom in the ring
and/or an S atom bound directly to the ring, and (2)
distilling the mixt. in the usual way.
   Compounds (II) = thiazole (T), 2,5-dimethyl T, benz
T, 2-mercapto T, 2-mercaptothiazoline, 1,2,3-thia-
diazole, benzothiadiazole, 2-thiohydantoin, 2-mercapto-
benzoxazole, 1,4-thiazine, methylene blue, 6-methyl-
2-thiouracil (III), etc.
   (I) = di, tri, tetra, or hexa-chlorosilanes, etc.
   Use in prodn. of pure silica for use in semiconductors. (26.9.69 as 6914639) (E13 DYNN)

22618R.   E11.   STACHE.20-6-66.
US66-558899.. R14.
STAUFFER CHEMICAL CO.   *GB-1187 497-R.
   C07f-07/04 (08-04-70)...
Improved production of hydrocarbonoxy silanes..   B5-.

The reaction may be represented as

$$SiX_4+4ROH \rightleftharpoons Si(OR)_4+4HX\!\uparrow$$

(where X is halo usually Cl, R is 1-8C alkyl, cycloalkyl or aryl).
   In order to remove the acid from the product and
increase the reaction rate the reaction is a) carried
out adiabatically in a solvent which will reduce the
solubility in the reaction mixture of the acid (b) react
the silane mixture in the presence of the solvent with
additional phenol or alcohol to the completely esterified
product. The process is particularly applicable to the
production of tetraethyl orthosilicate. (23.5.67 as
23781).

Fig. 8.6.   (Continued)

22940R.   A22-E12-G6.   HARINT.20-12-65.
US64-515143. . R14.
HARRIS INTERTYPE CORPORATION.
                            *CA-0836 467-S.
Sorkin-JL Thomas-DC. (sorkintho).
(10-03-70)..
Barrier layer for presensitised lithographic printing
plates..

Lithographic printing plate comprises a metal surfaced
support member and a hydrophilic barrier coating on
the support consisting of the reaction product of (I) an
OH-contng. water-dispersible urea-formaldehyde or
melamine-formaldehyde resin, and (II) a metal ester
of formula (II):-

$$(RO)_4M \qquad (II)$$

(where R is phenyl, tolyl, xylyl or a C_{1-8} aliphatic
group; and M is Ti or Zr).
Specifically the support member is pref. Al or Zn.
Compd. (II) is pref. tetraisopropyl titanate. A light
sensitive layer may be applied over the hydrophilic barrier coating. The ester (II) is applied in the form of a
soln. in an organic solvent, the soln. contng. sufficient
metal ester to deposit an amount of ester on the surface equivalent to that deposited by contacting the plate
with 0.1-10% wt. soln. for a period of time up to 10
mins. The ester on the plate is then hydrolysed and
reacted with the OH groups of resin (I) coating, to form
a hydrophilic reaction product substantially free of
oxides of M, and substantially non-susceptible to further hydrolysis. (9.7.66. as 965,071) (G5 HARI).

22987R,   E12.   GRACEW.22-04-69.
CA67-049375. . R14.
GRACE AND CO WR.
                            *CA-0836 608-S.
Godfrey-JJ. (godfreyjj).
(10-03-70)..
Preparation of sodium nitrilotriacetate...

Process for preparing sodium nitrilotriacetate by: (a)
reacting an aqueous formaldehyde solution with hexamethylenetetramine and HCN in the presence of a sulfonic acid to form an aqueous sulfonic acidnitrilotriacetonit-
rile mixture; (b) neutralizing the sulfonic acid and hydrolyzing the nitrilotriacetonitrile to sodium nitrilotriacetate with hot caustic soda solution; (c) cooling the hydroly-
zate to precipitate the sodium sulfonate and a portion of
the sodium nitrilotriacetate; (d) separating the mother
liquor from the precipitate; and treating said liquor with
caustic soda to precipitate substantially pure sodium
nitrilotriacetate therefrom; and (e) recovering the thus
precipitated sodium nitrilotriacetate. (22.4.69 as
049375)

23022R,   E12-K8-J4.   COMENE.19-09-63.
FR63-947959. . R14.
COMMISSARIAT A L'ENERGIE ATOMIQUE.
                            *DT-1443 663-R.
Auchapt-P Bagnols-C Bouzou-G. (aucbagbbu).
C07c C01g-57/00 (02-04-70)..
Continuous production of plutonium oxalate from its nit-
rate and oxalic acid. D6C.

Plutonium oxalate is prepared continuously by mixing a
soln. of plutonium nitrate with oxalic acid soln. at
room temp., separating the pptd. oxalate from the
mother liquor by decanting, followed by countercurr-
ent wading firstly with an acid soln. and then with water.
A part of the ppte., which is formed continuously,
is led to the decanting vessel, and then batch-wise to
acid wash column and the water wash column, while
the remainder of the ppte. between batches is recircul-
ated to the decanting vessel or to the acid wash column.
    Prior art processes (e.g. British Patent 882,950),
give plutonium oxalate which is not sufficiently pure
and has to be purified subsequently. The advantage of
the present method is that the mother liquor, which
contains impurities contained in the starting material,
is no longer mixed up with the wash soln. (9.9.64 as
C33843).

23044R,   E12.   REXCHE.25-10-61.
US61-147637. . R14.
REXOLIN CHEMICALS AB.
                            *DT-1518 110-R.
Singer-JJ Singer-JP. (singersin).
C07c (02-04-70).
Trialkalimetal nitrilotriacetate by continuous
saponification of nitrilotriacetonitrile,   D5-.

Cyclic process of production of crystallised trialkalimetal nitrilotriacetate (A) comprises: (I) mixing saponification solution containing (A) with aqs. solution of same
alkali metal hydroxide (B), at raised temp. (90-100°C),
(B) being used in excess of stoichiometric amount required for saponification of nitriloacetonitrile (C) in the
later step (IV); (II) mixture is cooled to 20-50°C; (III)
crystals of (A) formed are separated from the mother
liquor; (IV) mother liquor is treated with (C) and (V)
(C) is saponified at 90 - 100°C to (A).
    Good results are obtained by adding a 50% NaOH
solution to a 40% tri-Na-nitrilotriacetate solution in
such amounts,that 50 weight % NaOH solution are added.
(24.10.62 as R 33743)

23207R,   A14-E12-G6.   EASKOD.01-10-68.
US68-764330. (+01-10-68/US68-764332). R14.
EASTMAN KODAK COMPANY.
                            *BE-0739 708-Q.
Bass-JD Herz-AH. (bassjdher).
(16-03-70)..
Complexes of silver and organic compounds useful in
photographic development.   M3B.

Developing agent (I) for use in physical photographic
development of a latent image formed in a photosensitive metal salt composition, consists of a stable
organometallic complex, pref. of silver, which is
prepared from an organic compound selected from
thioamides, compds. contg. a guanyl group, mercapto-
acids, alkynes, hydroxyalkylcarboxylic acids, hydroxy-
heterocyclocarboxylic acids, cmpds. contg. thioalkyl
groups, oxalic acid, succinic acid, phenylenedioxy-
dialkylcarboxylic acids, and polymers contg. an atom
of the ligand type which can form a metallic complex
useful for physical development.
    The developing agent permits the advantageous
development of photosensitive compositions contg. a
dyestuff which has the property of forming active
development centres on exposure to light. Very
stable silver images are formed by the developing
agenc. (1.10.69). (A16 EAST).

Fig. 8.6. (Continued)

133

23261R.　　　A12-E12-G6.　　WESELE.03-10-68.
US68-764866. . R14.
WESTERN ELECTRIC COMPANY. *BE-0739 834-Q.
　Broyde-B. (broydeb--).
　(15-12-69).
Sensitizing additives for negative photoresists for
microminature circuits. M2B.

Pattern is produced on photoresist coated on substrate
by exposing parts to electron beam, eliminating those
parts not exposed and treating them by e.g. electro-
plating, etching or oxidation. Coating is a soluble resist rendered insoluble by exposure to electron beam
and contains compound enhancing sensitivity to electron beam by at least 10%, permitting the beam to
sweep the resist at a higher speed. The sensitizer
forms 50% of the resist coating. Preferably, photoresist is polyvinyl cinnamate or polymerised isoprene
dimers, and the sensitizer is an organometallic compound, soluble in same solvent as the resist, the organic radical being alkyl, aryl or aralkyl and the metal
from Group III, IV and V and atomic number >30.
　Improvement in sensitivity enables use of electron
beams, giving sharper imaging than optical methods.
(3.10.69). (A13 WELE)

23460R.　　　E12-L2-M22.　　MITSUB.25-10-66.
JA66-070320. . R14.
MITSUBISHI HEAVY INDUSTRIES LIMITED.
　　　　　　　　　　　*JA-7008 841-R.
　(31-03-70)..
Prepn of sand mould materials...

Ca salt of organic acid is added as a curing agent to
a composite sand or fluid sand for sand moulding.
　The Ca salt is, e.g Ca citrate, lactate or oxalate.
　Sand moulds prepd. from these materials have
sufficient strength after cure, and there is no fear
of explosion caused by evolution of gas; the fire-
resistance of sand mould materials is retained.
　The cured sand moulds are easy to break down
after casting because the Ca salt is decomposed by
heat of casting, and gas evolved in the decompn.
hinders the reaction between quartz sand and alkali
silicate. (25.10.66 as 70320).

23520R.　　　A18-E12.　　DNINKC.02-09-64.
JA64-049861. . R14.
DAI NIPPON INK AND CHEMICALS INC.
　　　　　　　　　　　*JA-7009 073-R.
　(01-04-70)..
Curing of vinyl type polymers containing an epoxy group
and/or hydroxy group..

Comprises using inner complex cpd. of carboxylic
acid esters in which Be, Mg, Ca, Sr, Ba, Zn, Cd, A,
Sc, Ti, V, Cr, Mn, Fe, Co, Ni, Cu or Zr is central
metal and carboxyl group or hydroxyl group is at β
position as curing ingredient.
　Cured material has excellent UV resistance, 3-D
structure and no discolouration. (2.9.64 as 49861/64)
(A21 DNIN)

31433F.　　　B3-C2-E12.
UPJOHN COMPANY.　　　　　　=GB-1186 938-R...
　C07d-27/04 C07d-29/12 (08-04-70).. (clg 26-9-66
US as 581730).
Aryl piperidino cycloalkane methanols as diuretics mo-
thproofers, pickling inhibitirs, herbicides, electrocard-
iograph jellies B3A.

Are new compounds of formula (I) including esters,
ethers, N-oxides, acid addition and quaternary
ammonium salts thereof:

(where n is 1-4; NZ is a 5-10 ring atom heterocycle;
R', R", and R''' are H, halogen, 1-6C alkyl or alkoxy
and CF₃; R₁, R₂ and R₃ are H, OH, CF₃, alkyl and
alkoxy).
　A preferred cpd. is cis-α, α-bis-(p-methoxyphenyl)-
2-piperidinocyclohexane methanol.
　The compounds are particularly useful as diuretics
and are prepared by reacting an appropriate cycloalkyl
aryl ketone (II):

with a Grignard reagent XMg—(/ ... R₁ (where X=I or Br).

(24.8.67 as 39082/67) (UPJO).

Fig. 8.6. *(Continued)*

GUIDELINES FOR THE PREPARATION
OF PATENT ABSTRACTS

Background

Rule 72(b) of the *Rules of Practice in Patent Cases* requires that a patent application include an abstract of the disclosure.

The patent abstract should enable a reader, regardless of his degree of familiarity with patent documents, to quickly ascertain the character of the subject matter covered by the technical disclosure and that which is new in the art to which the invention pertains.

The abstract is not intended or designed for use in interpreting the scope or meaning of the claims.

Content

A patent abstract should be a concise statement of the technical disclosure of the patent and that which is new in the art.

If the patent is of a basic nature, the entire technical disclosure may be new in the art, and the abstract should be directed to the entire disclosure.

If the patent is in the nature of an improvement in an old apparatus, process, product or composition, the abstract should include the technical disclosure of the improvement.

In certain patents, particularly those for compounds and compositions where the process for their making and/or use are not obvious, the abstract should set forth the process and/or use.

If the new technical disclosure involves modifications or alternatives, the abstract should mention by way of example the preferred modification or alternative.

The abstract should not refer to purported merits or speculative applications of the invention and should not compare the invention with the prior art.

Where applicable the abstract should include the following: (1) if a machine or apparatus, its organization and operation; (2) if an article, its method of making; (3) if a chemical compound, its identity and use; (4) if a mixture, its ingredients; (5) if a process, the steps. Extensive mechanical and design details of apparatus should not be given.

With regard particularly to chemical patents for compounds or compositions, the general nature should be given as well as the use; for example,

Fig. 8.7. Guidelines for the preparation of patent abstracts. *Source.* United States Department of Commerce, Patent Office, April 1969. Reprinted with permission.

"The compounds are of the class of alkyl benzene sulfonyl ureas, useful as oral anti-diabetics." Exemplification of a species could be illustrative of members of the class. For processes, the type reaction, reagents and process conditions should be stated, generally illustrated by a single example unless variations are necessary.

Language and Format

The abstract should be in narrative form and generally should be limited to one paragraph of 50 to 250 words. Since readers will use the abstract to determine whether the patent is within their special field of interest and whether the entire patent text should be obtained, the abstract should sufficiently describe the disclosure to permit readers to make this determination.

The language should be clear and concise and should not repeat information given in the title. Phrases should not be used which can be implied, such as, "This disclosure concerns," "The disclosure defined by this invention," and "This disclosure describes." The form and legal phraseology used in patent claims, often employing such words as "said" and "means," also should be avoided.

Responsibility

Preparation of the abstract is the responsibility of the applicant. Background knowledge of the art and an appreciation of the inventor's contribution to the art are most important in the preparation of the abstract. The review of the abstract for compliance with these guidelines, with any necessary editing and revision on allowance of the application, is the responsibility of the examiner.

Sample Abstracts

For further guidance, some sample abstracts are given below.

*   *   *   *   *

A heart valve with an annular valve body defining an orifice and having a plurality of struts forming a pair of cages on opposite sides of the orifice. A spherical closure member is captively held within the cages and is moved by blood flow between open and closed positions in check valve fashion. A slight leak or backflow is provided in the closed position by making the orifice slightly larger than the closure member. Blood flow is maximized in the open position of the valve by providing an inwardly convex contour on the orifice-defining surfaces of the body. An annular rib is formed in a

Fig. 8.7. *(Continued)*

channel around the periphery of the valve body to anchor a suture ring used to secure the valve within a heart.

*   *   *   *   *

A method for sealing, by application of heat, overlapping closure panels of a folding box made from paperboard having an extremely thin coating of moisture-proofing thermoplastic material on opposite surfaces. Heated air is directed at the surfaces to be bonded, the temperature of the air at the point of impact on the surfaces being above the char point of the board. The duration of application of heat is made so brief, by a corresponding high rate of advance of the boxes through the air stream, that the coating on the reverse side of the panels remains substantially nontacky. The bond is formed immediately after heating within a period of time for any one surface point less than the total time of exposure to heated air of that point. Under such conditions, the heat applied to soften the thermoplastic coating is dissipated after completion of the bond by absorption into the board acting as a heat sink without the need for cooling devices.

*   *   *   *   *

Amides are produced by reacting an ester of a carboxylic acid with an amine, using as catalyst an alkoxide of an alkali metal. The ester is first heated to at least 75° C. under a pressure of no more than 500 mm. of mercury to remove moisture and acid gases which would prevent the reaction, and then converted to an amide without heating to initiate the reaction.

Revised April 1969

Fig. 8.7. *(Continued)*

137

3,407,110
HEAT SHIELD
John S. Axelson, Levittown, and Calvin M. Dolan, King
of Prussia, Pa., assignors to General Electric Company,
a corporation of New York
Filed Nov. 25, 1964, Ser. No. 413,901
7 Claims. (Cl. 161—68)

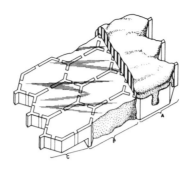

A flexible ablating heat shield, comprising a discon-
tinuous, heat-resistant, honeycomb matrix filled with a
flexible ablation material, is produced by (1) making cuts
in at least two of the cell walls of each of the honeycomb
cells of the honeycomb sheet to a depth less than the
thickness of the honeycomb cells but greater than the
thickness of the desired heat shield, (2) filling the cells
of the honeycomb with a flexible ablation material, such
as an elastomeric silicone, and (3) slicing the filled honeycomb perpendicular to the cell walls to remove that layer
of the honeycomb not cut in step (1) above.

3,406,514
**COMMUNICATION CABLE QUAD AND METHOD
OF MAKING SAME**
George Andrew Hanlon, Branford, and Albert James
Garde, Milford, Conn., assignors to The Whitney Blake
Company, New Haven, Conn.
Original application Jan. 27, 1966, Ser. No. 523,320, now
Patent No. 3,364,305, dated Jan. 16, 1968. Divided and
this application July 18, 1967, Ser. No. 654,156
7 Claims. (Cl. 57—156)

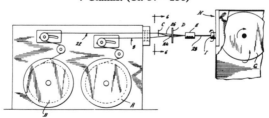

Disclosed herein is a quad cable construction made
from two joined wire pairs by slitting the web joining
one wire pair and then nesting the separate wires thus
formed in juxtaposition to one another along and across
the web of the other joined wire pair. The quad cable
construction is held intact by twisting thereof to entwine
the one joined wire pair with the two separate wires.

Quad cable thus formed exhibits minimal resistance un-
balance since the lengths of the pair wires cannot vary
and drawdown is minimized due to the joined condition
of the pair wires during processing.

Figure 8.8. Examples of abstracts of patents. *Source. Official Gazette of the United
States Patent Office.* Reprinted with permission.

3,406,573
CAPILLARY PIPETTE AND ADAPTER-
HOLDER THEREFOR
Robert C. Burke, Miami, Fla., assignor to Dade Reagents,
Inc., Miami, Fla., a corporation of Delaware
Filed Mar. 10, 1967, Ser. No. 622,210
10 Claims. (Cl. 73—425.6)

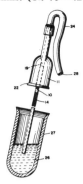

An adapter-holder to be used in combination with capillary pipettes. The adapter is provided with a central tapered bore which receives and frictionally grips the pipette, and the configuration of the adapter facilitates holding of the adapter and pipette in operative positions. The end of the adapter opposite the pipette-receiving end is of relatively small diameter and is received within a piece of flexible tubing. After the pipette is filled with the fluid to be tested, the fluid may be discharged and the pipette rinsed by alternately blowing and sucking on the tubing.

3,406,469
SLANT TOE FOOTBALL SHOE
Frank P. Dani, 46 S. Midland,
Mundelein, Ill.   60060
Filed Apr. 24, 1967, Ser. No. 633,196
4 Claims. (Cl. 36—2.5)

A football shoe having a square, flat toe at its front end and which is rearwardly inclined at its upper edge to about 20 degrees so as to permit a football to be kicked higher from placement and thus over a longer distance.

Fig. 8.8. *(Continued)*

3,407,261

ARRANGEMENT FOR THE FASTENING AND
CONTACTING OF A LAMINATED CONDUC-
TOR PLATE IN A SHIELD CASING

Hans Donath and Gerd Nothnagel, Munich, Germany,
assignors to Siemens Aktiengesellschaft, a corporation
of Germany

Filed Feb. 14, 1966, Ser. No. 527,344
Claims priority, application Germany, Feb. 16, 1965,
S 95,464
12 Claims. (Cl. 174—35)

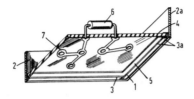

An arrangement for securing and supporting a circuit
board in a shielded casing using an intermediate frame
member. The circuit board has a metallized ground strip
along at least one edge. The frame member is soldered
to both the ground strip and the shield casing to electrically interconnect the ground to the casing. The frame member of one embodiment has an obliquely extending flange
contacting the wall of the casing. The angle of the flange
is the same as the angle of which the stream of flowing
solder is projected on the arrangement to facilitate the
soldering of the frame to both the ground strip and th'
casing.

Fig. 8.8. *(Continued)*

3,407,046
REACTOR FOR CONTINUOUS POLYMERIZATION
Jean-Marie Massoubre, Clermont-Ferrand, France, as-
signor to Compagnie Generale des Établissements
Michelin, raison sociale Michelin & Cie, Clermont-
Ferrand, Puy-de-Dome, France
Filed Jan. 27, 1966, Ser. No. 523,342
Claims priority, application France, Feb. 1, 1965,
3,889
3 Claims. (Cl. 23—285)

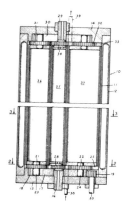

A reactor for the production of polymers is provided
with a hollow cylindrical shell and an eccentrically-
mounted carrier gear. Rotors oval in cross-section are
mounted on the carrier gear and simultaneously rotate and
gyrate to wipe the interior wall of the shell. The gyration
is about the axis of the carrier gear and eccentric with
respect to the axis of the shell.

Fig. 8.8. *(Continued)*

3,406,526
DOUBLE WALLED CRYOGENIC VESSEL
Dudley T. Lusk, Westmont, Ill., assignor to Chicago
Bridge & Iron Company, Oak Brook, Ill., a cor-
poration of Illinois
Filed Aug. 10, 1966, Ser. No. 571,565
7 Claims. (Cl. 62—50)

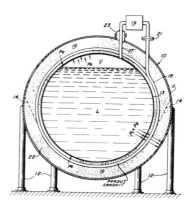

A double walled tank for the storage of cryogens where
the pressure in the ullage space in the inner tank is
counteracted by a pressure maintained in the space between the inner tank and the outer tank so that the inner
tank can be designed to withstand the forces created by
the weight of the stored liquid only without considering
the force created by the ullage pressure and designing the
outer tank to withstand the forces created by the weight
of the inner tank and the pressure in the annulus.

Fig. 8.8. *(Continued)*

3,406,606
DEVICE FOR THE TRANSPORT AND LAUNCHING
OF ROCKET PROJECTILES HAVING A RIGID
TAIL ASSEMBLY
Rainer Schöffl, Liebenau, Germany, assignor to Dynamit
Nobel Aktiengesellschaft, Troisdorf, Postfach, Germany
Filed July 11, 1966, Ser. No. 564,417
Claims priority, application Germany, July 15, 1965,
D 47,732
20 Claims. (Cl. 89—1.815)

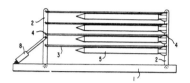

A frame that will hold rockets having rigid tail assemblies in a compact transport position where the rigid tail assemblies are transversely overlapping and is shiftable to a launching position where the rockets are held so that their rigid tail assemblies are transversely spaced from each other to allow clearance for independent firing of the rockets.

Fig. 8.8.   *(Continued)*

8.6.   ABSTRACTING FOR PERSONAL FILES

Many scientists maintain remarkably effective personal abstract files for their private use. They usually write the abstracts on 3×5 or 4×6 cards as they read and identify material of special interest. Since the material selected depends on the personal interests of the scientist, the content of the abstracts is highly oriented accordingly. Personal abstracts are usually very concise—sometimes telegraphic or coded in a kind of personal shorthand to save time. Obviously, no special rules need be followed for this kind of abstracting. The scientist should work in the manner most convenient to him. However, special care should be taken to identify carefully the original document so that it can be located again if needed.

If the abstracting is being done by members of a small research team working together, a few simple rules are highly desirable so that the team members can exchange abstracts and use one another's files. This situation is too variable to give any examples here.

More formal ground rules are necessary if personal abstract files are intended for later publication or for use by large numbers of individuals. These rules are as outlined throughout this text.

Proposed Front Page Patent Format

Request for Comments

The Patent Office proposes the adoption of an improved format for the front page of the patent document designed to incorporate the most pertinent information on a single page in an arrangement most suitable for efficient and effective utilization. The new format shall involve no change in the preparation or filing of an application by applicant.

As shown in the accompanying illustration, the information to be included comprises, when applicable, the following data elements: (The bracketed numbers relate to an International data element code.)

[11] Patent number
[45] Date patented
[21] Application number
[22] Filing date
[72] Inventor(s) (name and address)
[73] Assignee(s) (name and address)
Priority ([31] date, [33] country and [32] number)

[54] Title of invention
[52] U.S. classification (original and cross-references)
[51] International classification (one or more)
[50] Field of Search
[56] References cited and their classifications
Examiners' name(s)
Exemplary figure(s) of the drawing
Abstract of the disclosure
Exemplary claim(s) (if space permits)

Availability of the front page as a separate publication could bridge the gap which presently exists between the OFFICIAL GAZETTE announcement and the patent document. The proposed format is also intended for utilization as a "first-line" searching expedient.

The new front page would be printed by computer photocomposition in conjunction with the Patent Office program for establishing a computer supported library of patent documents in machinable form. Only those patents which are printed by computer photocomposition will include a front page format of the type proposed. It is anticipated that this program will begin in June 1969 with a small portion of the weekly patent issue followed by a rapid build-up thereafter. During the transition period, the remaining patents will be printed by the conventional method in the current format. In the first year of operation, about 50% or 35,000 of the allowed applications would be converted to data bases and printed by computer photocomposition. Expectations are that the total patent issue would be processed in this manner by July 1971. It is presently contemplated that any additional issue fee for the new front page would be disregarded as provided under 35 U.S.C. 151.

All persons who desire to present their views, objections, recommendations in connection with the proposed format are invited to do so by forwarding the same to the Commissioner of Patents, Washington, D.C., 20230, on or before April 15, 1969. No hearing will be scheduled.

EDWARD J. BRENNER,
Commissioner of Patents.

Jan. 17, 1969.

(Format on next page.)

Fig. 8.9. Proposed front page patent format. *Source. Official Gazette of the United States Patent Office,* March 11, 1969. Reprinted with permission.

144

[19] United States Patent   [11] 3,339,457

[45] Patented Sept. 5, 1967
[21] Application No. 465,394
[22] Filed June 21, 1965
[72] Inventor: L. Pun, Geneva, Switzerland
[73] Assignee: Brevets Aero-Meccaniques S.A.,
          Geneva, Switzerland
          a Society of Switzerland
[31] Priority: June 26, 1964, Aug. 18, 1964
[33]   Luxembourg
[32]   Nos. 46,404 and 46,787

[54] FIRE CONTROL SYSTEMS

[52] U.S. Cl. 89/41, 235/61.5
[51] Int. Cl. F41g 5/00
[50] Field of Search 89/41, 235/61.5

[56]   References Cited
     2,710,720 ......U.S. ...... 6/1955.................235/61.5
     2,922,572 ......U.S. ...... 1/1960.................235/61.5
     2,977,049 ......U.S. ...... 3/1961...............235/61.5 X

Primary Examiner— B. A. Borchelt
Assistant Examiner— W. C. Roch

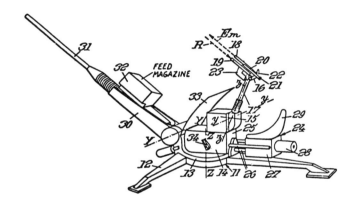

ABSTRACT

The angular coordinates of a target are determined continuously whereas the distance thereof is determined at regular time intervals. The angular coordinates of the gun are calculated from said angular coordinates of the target, said distance, the duration of said intervals and the firing tables, the time of flight of the projectile being calculated by an algorithm, repetitive, but without iteration, from an estimated time of flight and said duration. The target is preferably followed by a laser rangefinder.

CLAIM

2. A fire control system for directing a gun toward a target which comprises, in combination,

means for determining substantially in a continuous manner the bearing and the site of said target,

means for determining the distance of the target from the gun at regular time intervals, and

means for calculating the angular coordinates to be given to said gun so that a projectile fired therefrom normally reaches said target, from said angular coordinates of the target, the distance of the target from the gun, the duration of said time intervals and the indications of firing tables giving, for said distance, the time of flight of the projectile and the elevation angle to be given to the gun, the calculated time of flight of the projectile being determined, by an algorithm, repetitive, but without iteration, from an estimated approximate time of flight of the projectile and from the duration of said time intervals.

Fig. 8.9.   (Continued)

8.7. ABSTRACTING CLASSIFIED DOCUMENTS

A government confidential, secret, or other "classified" document is one which contains information that is important to the security of the nation. There are some documents that are so confidential that even the abstract must be treated as classified, and in any event, all classified documents must be handled in accordance with security regulations by an abstractor with proper security clearance. However, if it is decided that a classified document can be announced in an unclassified abstracting bulletin, an unclassified abstract (one with no confidential data) must be prepared.

The preparation of an unclassified abstract of a classified document is not as difficult as it may appear at first. One approach involves the extensive use of sanctioned abbreviations, and especially of code names and code numbers in referring to products and processes studied. This usually conceals identity from "outsiders," but for individuals with the "need-to-know" and security clearance it may provide sufficient information to warrant the reader's requesting the full original document. Also, an unclassified abstract of a classified document will often be briefer and less informative in nature than other kinds of abstracts. Thus, the abstractor will probably omit detailed quantitative data on speed (and other performance characteristics) and cost. For example:

> Current work on Project Hamden is now centered on alternates 23A and 23B. The results of the experiments completed so far show an increase in efficiency of more than 50 percent.

Techniques such as these enable the abstractor to write an unclassified abstract and still convey useful information to the reader who has the security clearance and the "need-to-know" as specified by security regulations.

Suggestions for "Abstracting Scientific and Technical Reports of Government Sponsored RDT&E" will be found in Document AD-667,000, U. S. Defense Documentation Center (see Appendix). The abstractor will want to make special note of the standard "work form" on which the abstract is to be recorded.

chapter 9
Providing Access to Abstracts: Grouping and Indexing

It is not enough to write good abstracts. If abstracts are to be used, they must be accessible.

9.1. GROUPING RELATED ABSTRACTS

In all but the most modest of abstract efforts, it is customary to aid the abstract user by the simple expedient of grouping related abstracts. Thus, an abstract bulletin devoted to a specific product might have the following sections:

1. Research and development.
2. Manufacturing processes.
3. Uses.
4. Marketing.
5. Safety aspects.

There are many other ways of grouping abstracts, such as by chemical nature, biological species, or Standard Industrial Classification Numbers (13). If the method is carefully selected, and if the abstracts are properly placed, the abstractor can save a great deal of time for the user.

Since abstracts can often be placed in more than one section, it is customary to cross-reference such abstracts in a fashion such as the following:

MANUFACTURING PROCESSES
    see also Abstract Nos. 10, 16, 25, and 84.

Also, the abstractor may want to refer to related or equivalent work previously abstracted. He does not abstract equivalent work again (e.g., the United States equivalent of a previously abstracted British patent) but instead cites the United States patent and refers the reader to the abstract of the British patent by simply citing the latter's number and abstract reference.

The numbering of individual abstracts is an aid when used as just described and is also helpful in the indexing process.

9.2. THE INDEXER AND ABSTRACTING

The use of abstracts for indexing purposes is so important and so widespread that it deserves special emphasis. There are important implications for abstract user, abstractor, and indexer.

For the user, good indexing means success in locating pertinent abstracts and access to vital information. The abstractor and indexer need to be constantly aware of this and must work together as closely as possible.

Also, abstracts are an important part of many information storage and retrieval systems and as such are intimately related to indexing quality and system performance.

Indexing from the full original document is, in theory, the route of choice. If the original is not available, however, the product of the abstractor assumes special importance. In any event, whoever prepares the abstract should try to make it as rich in indexable information as possible. Indexing of abstracts can best be done from full, high quality abstracts.

As we have said, the indexer is a vital cog in the abstracting process since most scientists rely almost completely on indexes for the location of abstracts. For this and other reasons, it is important that indexing of abstracts be accurate, thorough, and consistent.

The indexer can be more consistent if he uses a thesaurus, which specifies acceptable index terms and also delineates cross references as shown in Fig. 9.1. The thesaurus is of particular help in handling synonyms.

Thoroughness deserves special emphasis. What we mean by this is depth of indexing or number of index entries per abstract. The chances of the user locating an abstract are enhanced by deep (thorough) indexing. Some large published services average only about 6 subject index entries per abstract. Because of the limitations of language as a means of communication, shallow indexing can make index use a guessing game at which the index user often loses. The seeker of generic information or of concepts for which there are many synonyms is often frustrated by shallow indexing.

In theory, a minimum of limitations should be placed on indexing depth.

Control rods
Reactor cores
Reactor fuel cladding
Reactor fuel plates

Reactor Fuels
(Nuclear Reactor Technology)
Specific to:
  Reactor materials

Reactor Hazards
(Nuclear Reactor Technology)
Specific to:
  Hazards
  Reactor operation
Also see:
  Confinement
  Containment
  Reactor accidents
  Reactor inspection
  Reactor safety systems

Reactor Inspection
(Nuclear Reactor Technology)
Specific to:
  Reactor operation

Reactor Kinetics
(Nuclear Reactor Technology)
Includes:
  Chain reactions
Specific to:
  Reactor operation
Also see:
  Reactor control
  Reactor theory

Reactor Lattice Parameters
(Nuclear Reactor Technology)
Includes:
  Migration area
  Migration length
Generic to:
  Buckling (neutron density)
  Conversion ratio
  Fast fission factor
  Fuel burn up
  Multiplication factor
  Neutron age
  Neutron flux
  Neutron lifetime
  Non-leakage probability
  Power coefficient of reactivity
  Reactor power density

Reactor reactivity
Resonance escape probability
Temperature coefficient of reactivity
Thermal utilization
Also see:
  Fuel enrichment
  Reactor fuel processing
  Reactor operation
  Reactor system components
  Reactor theory

Reactor Materials
(Nuclear Reactor Technology)
Generic to:
  Burnable poisons
  Reactor coolants
  Reactor fuel cladding
  Reactor fuels
  Reactor moderators
  Reactor shielding materials

Reactor Moderators
(Nuclear Reactor Technology)
Specific to:
  Reactor materials
Also see:
  Neutron absorbers

Reactor Operation
(Nuclear Reactor Technology)
Generic to:
  Confinement
  Containment
  Reactor accidents
  Reactor control
  Reactor feasibility studies

  Reactor hazards
  Reactor inspection
  Reactor kinetics
  Reactor shutdown
Also see:
  Reactor theory

Reactor Poisons
(Nuclear Reactor Technology)
Specific to:
  Neutron absorbers
Generic to:
  Burnable poisons
  Fission product poisoning
Also see:
  Reactor materials
  Reactor moderators

Fig. 9.1. Specimen page from the *Thesaurus of ASTIA Descriptors. Source.* F. W. Lancaster, *Information Retrieval Systems,* p. 37, Wiley, New York, 1968. Reprinted with permission.

Even if it takes 50 to 100 entries to index thoroughly, this effort usually costs substantially less than the value of information thereby made accessible. But many organizations find that practical considerations of budget make it necessary to limit indexing efforts.

As will be noted in Chapter 14, the use of computer-based information input and retrieval systems can make the full text of an original document or abstract accessible and can obviate the need for human indexing in some cases.

Indexing and other activities of information retrieval as looked at by the evaluator of information systems are treated by Lancaster (15). Figure 9.2, taken from Lancaster, shows the activities of "information retrieval" with emphasis on the role of the indexer. In that figure, abstracts can be regarded as a form of input although not shown as such in the figure.

9.3. ABSTRACTING AND INDEXING BY ONE PERSON

It is important to note that often both the abstract and the index entries may be prepared by one individual. Indeed, some think that this is an ideal situation because after writing an abstract an individual is theoretically in an excellent position to prepare appropriate index entries. This is more likely to take place in relatively small, in-house operations, but it can also take place in very large organizations.

9.4. ABSTRACTING AND INDEXING BY DIFFERENT INDIVIDUALS

In some cases, especially in larger organizations, the abstracting and indexing steps are performed by different individuals.

When abstractors and indexers are in the same organization, and especially in the same building, a mutually satisfactory working relationship can be developed. The indexer will know when and how he can depend on the abstractor for his index entries; he may even ask the abstractor to suggest index entries. He will also know when he must look at the full original document—for example, this is very often the case with the indexing of papers on synthetic and theoretical organic and inorganic chemistry. Index entries to information that is not found in the abstract are likely to appear in such cases.

Sometimes the writer of the original document is in the same organization or even the same building as the abstractor and indexer. Or he may be under contract, as with work done on government projects. In such instances, not only can the writer of the original be asked to supply an

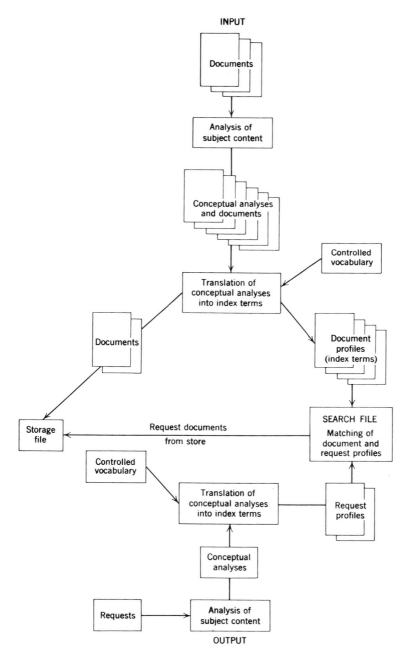

Fig. 9.2. The activities of "information retrieval." *Source.* F. W. Lancaster, *Information Retrieval Systems,* p. 4, Wiley, New York, 1968. Reprinted with permission.

author abstract fully suitable for indexing, but he can also be asked to suggest so-called key words for indexing purposes.

In any event, the abstractor should endeavor to find out whether his abstracts are likely to be indexed by someone else, either now or in the future, and if so, how and by whom.

chapter 10
In-House Abstracting Bulletins

10.1. INTRODUCTION

Many organizations publish in-house abstracting bulletins that cover current literature with the basic purpose of keeping their personnel up to date. But because the in-house abstracting bulletin is a venture that can involve many thousands of hours, and thousands of dollars, per year, it should not be undertaken lightly. A full scale, whole-hearted commitment of people and funds is necessary to achieve true success.

If the basic purpose is to be achieved, the in-house bulletin must be, above all, timely. It must also be presentable, legible, readable, and generally palatable. Relatively small matters can make a big difference. For example, overly thick bulletins (more than 30 to 40 pages per week) may discourage reading, as may those bulletins which are miniaturized with small type (or via photoreduction techniques) or those which are printed on very shoddy paper. Most of the subject matter in the bulletin must be relevant to the broad interests of the audience served; frequent irrelevance is costly, time-consuming, and generally wasteful.

Many of the comments in this chapter will also pertain to small, specialized, externally published services.

10.2. ADVANTAGES

In-house abstract bulletins, serving as they do the scientists of a relatively small private or public organization, can have many tangible benefits, if properly managed. One of the principal advantages is speed. With a properly managed effort, in-house bulletins can abstract many publications much more quickly than other kinds of abstracting services.

One reason for this is that in-house bulletins are usually relatively small, and produced completely on-site. This usually means that the production cycle can be controlled; also, internal bulletins are not subject to mail delays if the audience is all in one location.

Another advantage of in-house bulletins is that they are selective. Although the careful bulletin editor will want to include some peripheral material to stimulate ideas and to keep his readers from becoming inbred and narrow-minded, he will usually limit his coverage to the broad scope of his organization's interests.

And the editor can orient the content of his abstracts toward his organization's interests—if these abstracts are prepared internally in the first place. On the other hand, if the editor uses published abstracts (clipped out) in the bulletin, such orienting becomes difficult but not impossible. It can be partially achieved by such means as underlining for emphasis or the addition of a typewritten note at the end of the clipping.

In-house bulletins also make it possible to cover a wide range of both conventional and nonconventional documents, some of which are beyond the scope of external bulletins. Business data, trade literature, press releases, internal documents, and preprints are examples of nonconventional materials that might not ordinarily be covered in external science and technology abstracting services. Selective coverage of meeting paper abstracts is also possible; *this is important because a high percentage (perhaps half) of such papers are never published.*

The inherent constraints of published services, especially the larger services, offer other opportunities and advantages to the editor of an in-house bulletin. For example, some of the larger services will publish only information that is novel as, for example, on new chemical compounds, and there may be requirements that only clearly identified species or compounds be covered. Also, the results of scientific research are emphasized while material which is less scientific, such as business (economic) data or perhaps techniques (such as how to grow peas) are often deemphasized by the external services.

We mention these constraints not as criticism, but to indicate that the volume of literature being published requires that even the largest service must draw the line somewhere.

On the other hand, the in-house or specialized service can cover material such as mentioned above *and can do so in depth and detail* because of a much more limited scope.

10.3. DISADVANTAGES

The principal disadvantages of the in-house bulletin are, as mentioned earlier, extensive investment in time and funds and also, paradoxically, limited scope.

10.4. DECIDING ON AN INTERNAL BULLETIN; ALTERNATES

At the very outset, management must ascertain needs and then decide whether an already available externally published abstract bulletin will do the job with reasonable economy and speed. *Often, an existing external service will be found to be adequate.*

If it is decided that no existing service will do the kind of job that is needed, before committing an organization to an in-house abstract bulletin, management should consider one or more of the following *alternates* or *supplements:*

1.   Circulate tables of contents of current journals by such means as a subscription to one of the *Current Contents* publications of the Institute for Scientific Information. These kinds of services are fast and relatively economical. Another example of this kind of service is *Current Papers on Computers & Control* published jointly by the Institution of Electrical Engineers (London) and the Institute of Electrical and Electronics Engineers (New York); this is a categorized list of citations to various kinds of literature in the fields covered.

2.   Selectively disseminate information. Pinpoint it to specific needs as compared to the "shotgun" approach of the abstract journal. Let us describe Selective Dissemination of Information (SDI) in a little more detail for the reader who may not be familiar with it. This concept usually means the dissemination of information by information center personnel to individuals who have previously indicated their specific interests. This hopefully means that individuals will receive only information of special interest to them. Information can be received in the form of titles, abstracts, or even complete documents depending on the system used.

SDI systems can be computer-based systems, or this kind of service can also be conducted in a completely effective manner using a *manual* system. For example, the Technical Information staff at Olin Corporation, in New Haven, Connecticut, in one phase of their operations disseminates manually about 200 current alerting notices each month, with about 90 percent relevancy based on replies; this is in addition to the use of computer systems. SDI is more fully discussed in Section 14.4.

3.   Establish "journal clubs." Under this plan each scientist is assigned

a small group of journals to read and to report on to colleagues in his "club" at regular intervals.

4. As a matter of formal policy, encourage all scientists in the organization to spend several hours per week on the job keeping up with current literature.

5. As a matter of policy, encourage scientists to attend in-house seminars and communications meetings designed to inform them of current developments. This could include an in-house continuing education program.

10.5. THE BULLETIN EDITOR

If it is decided to have an internal bulletin, the next step is the selection of a well-qualified editor. The editor should be thoroughly familiar with the interests of his organization as a whole and with the specific needs of its personnel—both present needs and anticipated needs. He can work more effectively if he is aided in his job by laboratory personnel who can make suggestions as to input, and he should not hesitate to call on them for suggestions as to his product. Also, if he works in a technical information center, his colleagues there should offer suggestions. As mentioned later, the editor should seek continuous feedback from all who get his bulletins.

10.6. ORGANIZATION OF IN-HOUSE BULLETINS

In-house bulletins can be of the omnibus type in that different kinds of literature and a variety of fields are intermixed in a single bulletin. But a more practical approach is for the editor to organize his bulletin into broad categories based on the interests of the people in the organization. The reader can then elect to receive only those categories of special interest to him. (This also saves on printing and distribution costs.) Another option is to have separate bulletins for different kinds of literature, such as patents, journal articles, and internal reports. A simple table of contents in each issue is essential.

In the production of in-house bulletins, it is highly desirable to number the abstracts either consecutively or in some other definite way for ease of reference. This is particularly desirable and even necessary if the bulletin is indexed.

It is essential to describe to the audience exactly what each of the sections of the bulletin covers—that is, to define the scope of the parts of the bulletin so that recipients will know what to expect and also to ensure that they will get only those parts they want and need. An aid in this direction is to indicate the list of journals and other sources covered in each of the parts. This also enables readers to suggest additional sources to be covered, or deletions.

10.7. KINDS OF IN-HOUSE BULLETINS

10.7.1. In-House Bulletins—Patent

Bulletins covering United States patents are relatively easy to prepare. A workable technique is to have the bulletin editor select pertinent abstracts from the *Official Gazette of the United States Patent Office,* and then to have a clerk clip the abstracts directly from the *Gazette.* The clerk then arranges and pastes up the abstracts (4 to 6 per page) according to the sequence indicated by the editor—usually an arrangement by field.

The whole process is straightforward but time-consuming. For example, selection of 12,000 patents per year for an in-house bulletin can require the editor to spend about 8 hours per week, and the clerk about 24 hours per week. In addition, mimeographing and distribution costs can be extensive; in fact, the mimeographing costs alone can make up the major fraction of the total cost.

Subject and author indexes to patent abstract bulletins are very helpful, but these are an additional cost factor and are eventually made obsolete by conventional externally published indexes. Provision of a simple table of contents showing the categories of subject matter in each issue of the bulletin is highly desirable, and it can be produced at minimal cost.

Inclusion of abstracts of foreign patents is highly desirable if available resources make this possible. Foreign patents are important because they often constitute the first public disclosure of important inventions which may become commercially significant. Such abstracts appear in foreign patent gazettes, in specialized patent-abstracting publications, and often in more generalized abstracting publications. A "right-to-copy" or royalty fee may need to be paid if the abstracts are clipped out and reprinted. This fee applies especially to copyrighted services, but does not apply to many government services which are not copyrighted or otherwise limited.

In some cases, of course, the editor may want to write his own patent abstracts (see Section 8.5) for either United States or foreign patents. This makes orienting possible but can consume substantial amounts of time.

10.7.2. In-House Bulletins—Journal Literature

This kind of bulletin is not easy to produce, since the amount of potential input is much greater than for patents and can come from a great variety of sources. Limits must be set.

One favored approach is for the editor to work from a carefully selected list of key journals. These, plus suggestions from his colleagues, should give him an adequate data base from which to make selections. Speed of input can be increased if author abstracts are used. The editor himself can prepare brief indicative abstracts if author abstracts are not available.

10.7.3. Newsletters

Another related approach toward a form of in-house bulletin is a periodic (monthly) scientific highlights newsletter. Coordination of this effort is best handled through a specific individual in the Technical Information Center. This requires both critical selection and abstracting by in-house laboratory experts in specific fields. In theory, at least, it requires no extra work from the expert—he should be doing this anyway. The aim is to produce a publication which will inform people about broadly significant new developments.

The production of a bulletin of this kind requires, first of all, the selection of experts within the organization in various fields and then, in turn, the selection by these experts of what is significant in the course of their personal reading. This depends largely on the inherent value and novelty of the work as judged by the abstractor.

Once the expert abstractor makes his selections, he describes the reported work briefly and indicates why he considers it significant. For example, see Fig. 10.1.

A bulletin of this type should be very brief (a page or two), issued promptly, and prepared with great care so that people will read it.

First Stereospecific Synthesis of the C-18 Cecropia Juvenile Hormone
                An incredibly elegant synthesis of this hormone, featuring
1) the stereospecific conversion of propargylic alcohols to trisubstituted
olefinic carbinols via organoaluminum and copper reagents, 2) selective
propynylation, 3) the one-flask conversion of allylic alcohols to con-
jugated esters, and 4) the two-step homologation of primary halides,
has just been reported by E. J. Corey (JACS, 5618 (1968)).

Azobenzene Complex of Divalent Palladium

The reaction of $[(C_{12}H_9N_2)Cl]_2Pd_2$ with four moles of triethylphosphine
results in the formation of trans-Pd $[P(C_2H_5)_3]_2(C_{12}H_9N_2)Cl$. Perhaps
the most interesting aspect of this structure is that the potentially
bidentate ligand, azobenzene, forms, instead, a stable, single σ-bond
to the metal. The crystallographic structure of this complex is reported.
Chem. Comm., 17 (1021 (1968)

Fig. 10.1.   Examples of abstracts from a scientific highlights newsletter.

10.8. EVALUATION

Once an in-house bulletin is established, both the editor and management should constantly scrutinize its value. A fairly simple questionnaire to all recipients issued at intervals of 6 to 12 months will indicate whether recipients want to continue to receive the bulletin and will help remove from the circulation list the names of those who no longer want the bulletin. The questionnaire should also help elicit comments on needed improvements and on the value of the bulletin.

Bulletin editors and management should be willing to discontinue bulletins which few people like, need, or read. Too often there is a tendency to continue publishing a bulletin which no longer meets real needs. It takes a keen sense of what is and what is not truly important to the organization and its personnel to make this decision.

Many organizations impose an internal accounting charge for bulletins and related services. This charge can be remarkably effective in identifying those who show that they really want a service by agreeing to pay for it.

10.9. BULLETIN BY-PRODUCTS

The same abstracts that appear in the in-house bulletin can serve other uses. The imaginative editor will try to get as much mileage as possible from this bulletin.

One option, for example, is to print the abstracts so that these can be cut up by the individual scientists for use in their personal files. This requires first of all that the abstracts be printed on one side of the page only. Another help to the scientist is if the abstracts are made to fit within a preoutlined 3×5 or 4×6 card format for ease of cutting up and filing.

By use of mimeographing techniques, the same input can be used to print directly onto 3×5 catalog cards for use by both information center and laboratory personnel.

Still other options are open if the input is "machine readable"—understandable by the computer. These are discussed in Chapter 14.

10.10. BULLETIN PRODUCTION: SAVING TIME AND FUNDS— A PARTIAL RECAPITULATION

We have already mentioned some of the techniques that the abstract bulletin editor can use to speed up his work and to lower costs. For convenience, we recapitulate some of these:

1.   Use dictating machines for speed.

2. In conjunction with point 1 above, use electronic devices to "capture" typing to help avoid repetitive typing.

3. Use valid language shortcuts.

4. Use author abstracts to the maximum extent.

We have previously mentioned, but should mention again, the value of dividing the abstract bulletin into parts as a factor in cost reduction.

There are other related ways to save time and funds. For example, we have mentioned that it may be desirable to cut out and paste up abstracts from external abstracting services in a logical sequence. This not only saves time, but eliminates some errors in the transcription process.

In the case of *Chemical Abstracts,* lease of the microfilm edition covers the fee and the permission necessary for copying full abstracts for in-house use *under specified conditions.* (If an organization does not get the microfilm edition, a special right-to-copy fee must be paid.)

The editor can orient or modify these "ready-made" abstracts by such simple techniques as:

1. Underlining to indicate key words, phrases, authors, or organizations.

2. Adding an explanatory phrase to indicate special relevance.

3. Referring to related work previously reported.

Reproduction from the original document of graphic material (such as chemical structures, summary tables, or diagrams of mechanical and electrical devices) also reduces costs. Retyping produces errors, and expressing graphic material in words is often cumbersome or even impossible—hence the editor tries to avoid these.

Such techniques allow the abstract bulletin editor to concentrate on materials not abstracted elsewhere, or which may deserve special attention because of the organization's special interests.

Another technique for controlling costs is to print the in-house abstract bulletin on economical (but not shoddy) paper. Also, printing can be "back-to-back" (i.e., with printing on both sides of the page). This may inconvenience some scientists who like to cut out abstracts for their personal files, but it is rarely possible to please everyone, no matter what format is used. Some organizations try to solve this problem by sending two copies of the bulletin to some recipients.

But the major factor in keeping in-house bulletin costs at a reasonable level is to control the scope and size of the bulletin; in other words, the number of abstracts. An overly ambitious scope can lead to monumental production logjams. The editor is, therefore, highly selective. He is aware that he may be criticized for omitting material, but he replies to such criticisms with these points:

1.   The individual scientist has an inescapably *personal* responsibility for doing his best to keep up with literature in his field. No amount of work by the editor can cancel out this responsibility.

2.   There *are* comprehensive published bulletins. The fundamental goal of the *in-house* bulletin editor is to be *selective*.

10.11. DISTRIBUTION

The editor will want to be sure that management approves the names of those who will receive the bulletin—although in many cases the responsibility will be redelegated back to the editor. Controlled distribution is important for at least three reasons: First, the content of a bulletin can often tell much about an organization's interests; hence, it is often regarded as a proprietary document. Second, the bulletin should go only to those who are likely to make good use of it; this precludes, for example, distribution to most nontechnical personnel. Third, controlled distribution means fewer copies and, therefore, lower costs.

The evaluation process described earlier in this chapter ensures that the distribution list is up-to-date and truly responsive to real needs.

chapter 11

The Role of the Abstractor in Literature Searching

As already mentioned, there are cases in which the abstractor must be concerned not only with the individual abstract but also with many abstracts in relationship to one another. For example, in literature searches and critical reviews, all of the abstracts (or other material) together must form a coherent, well organized unit.

We shall here discuss an abstractor's approach to in-house literature searching. The function of literature searching is one performed frequently by those abstractors who also function as information scientists; this is an especially widespread practice in industry.

Abstractors perform literature searches for laboratory scientists to avoid unnecessary duplication of work, to build on a firm foundation of facts, and to spark new ideas.

11.1. SCOPE OF THE SEARCH

In writing abstracts for literature searches, the abstractor usually has the advantage of a personal working relationship with the client. The abstractor and the client should have a careful interview, and should agree on the scope of the search before any work is done by the abstractor.

It is best to put the client–abstractor agreement in writing, rather than to leave it in oral form, so that there can be no misunderstanding. This agreement should include such information as the reason the search is needed, what will be done with the results, the specific aspect of interest, and the time span to be covered. Exclusions and limitations should be spelled out with particular clarity. It is also helpful to know how soon the search must be completed, and whether there is an upper limit or ceiling on the funds that will be spent in the course of the search.

For example, a client may request a search on the geology of a specific part of Alaska in connection with studies intended to help locate oil deposits. He may further specify that he is interested only in articles published in the United States from 1960 on. He needs the results within 60 days after the request is made. A search request form as shown in Fig. 11.1 will help develop the information needed by the abstractor.

11.2. ALTERNATIVES IN REPORTING RESULTS

The agreement is restated by the abstractor in his written report to the client. The scope and objectives, and any conclusions reached, are spelled out just as in reporting about laboratory research. Sources used and references to any previous related searches are cited to avoid redoing work already done.

The needs of the client for speed and for depth of approach are key factors in selecting the kind of report to be written. The client should be presented with a choice of alternatives—a choice that depends on the client's particular needs, on the nature and magnitude of the field to be searched, on available resources, and on abstractor capabilities. The alternatives should be clearly identified during the course of the abstractor-client interview.

The alternatives which the abstractor may want to present to his client include the following:

1.   Nonnarrative reports carefully organized into report form, consisting of (*a*) copies of abstracts from externally published services and/or (*b*) individually written abstracts tailored to the specific needs of the specific client.

2.   Narrative reports which may or may not consist of critical reviews.

In the *nonnarrative* approach, the abstractor identifies pertinent documents and then collects or writes individual abstracts. He arranges these carefully either by subject or chronologically, with separate groupings by the type of document (patent, article, trade literature, or other). It is quite common, and even desirable, to have a page or two of introductory remarks prior to the presentation of the abstracts; this is the only narrative material in this approach.

Indexes to organizations, authors, subjects, and patent numbers (if the search includes patents) are highly desirable when this approach is used.

REQUEST FOR TECHNICAL INFORMATION   No. _____
            Please send to: Manager, Technical Information Services

Requested by   Tel. Ext.   Location   Charge No.
                                    Billing Code

Authorized by   Date of Request   Date Needed

Specific Subject
(Please identify subject clearly. Explain why information is needed and how it will
be used.)

Scope of Subject: Use   Preparation   Manufacture   Marketing

                    Other Areas
                    (explain)

                              SCOPE OF SEARCH
Time span to be covered   Material
                            to be   Patents   Non-Patent   Both
                            covered   only   material only

Other limitations such as by country, language

Dollar limit (cost of search)

Sources already checked by originator

Sources recommended by originator

Companies, laboratories or research workers believed active in this subject

   May this search be done in part or whole by outside searching service?   Yes   No

To be reported: (Check one)
                    Submit when   Advance draft copy   As pertinent info
                    finished   for speed   is located

In form of: Specific facts   Titles and   Abstracts   No. of copies of
                                 References only   report needed

              Other   Reprints of articles   CA Indexes
              (explain)

Continuous Awareness after report is finished?   Yes   No

                        Do Not Write Below This Line
Received by   Date Received   Date Reported

Assigned to: Acknowledged in writing by: Date Acknowledged:

Fig. 11.1.   Literature search request form.

As in the case of in-house bulletins, the abstractor can save much time by first selecting and sequencing and then having a clerk copy, clip, and paste up relevant abstracts from external abstracting services.

The abstractor will frequently want to exercise the options of (*a*) "editing" published abstracts by adding pertinent comments, or by underlining key words, and (*b*) writing his own if externally published abstracts do not exist or are inadequate to meet the specific need. The second option gives him an especially good opportunity to relate the abstracts to one another, such as by emphasizing similarities and differences between the materials abstracted and by placing special emphasis on what is most important.

In the *narrative* approach, abstracts are not used; instead the salient features of the documents previously identified as pertinent are woven together in a narrative framework. The arrangement of material is by broad categories of subject matter logically arranged. The material is "self-indexed" if this kind of arrangement is followed. The references are usually grouped at the end of the report and are referred to throughout the text by reference number.

The narrative approach is more difficult and takes longer to prepare than the nonnarrative approach. It lends itself well to critical interpretation of findings—although this requires even more time and also assumes that the writer is highly competent in the field being reviewed. The writer of a critical review intended for publication will find helpful suggestions in a paper by Harold Hart (16).

11.3. PRESENTATION OF DATA AND FINDINGS
IN TABULAR FORM

The prudent use of tables in searches can save time for the abstractor, typist, and reader. Another advantage of tables is that they can be quickly and easily understood by impatient scientists who prefer "digested" information presented in tabular form.

Tables are especially well suited for covering a large number of references in highly condensed form. For example, if a large number of chemical compounds have been made by one general reaction, the abstractor can conveniently summarize these in tables with reference numbers to the original documents. (See Fig. 11.2.)

There is at least one major disadvantage connected with the use of tables: the amount of information that can be presented about each reference is limited.

Fig. 11.2. Example of tabular presentation. *Source.* O. C. Musgrave, "The Oxidation of Alkyl Aryl Ethers," *Chemical Reviews*, **69**, No. 4, August 1969, p. 502. Copyright 1969, published by American Chemical Society. Reprinted with permission of copyright owner.

Table I

Scholl and Related Reactions

Ether	Reagent	Products	Ref
Anisole	γ irradiation in presence of benzoyl chloride	2,2′, 4,4′, and possibly 3,3′-dimethoxybiphenyl in very low yield	33
Anisole	Thermal neutron irradiation	4,4′-Dimethoxybiphenyl (low yield)	34
Anisole	Benzoyl peroxide and aluminum chloride in nitrobenzene	4,4′-Dimethoxybiphenyl (23%)	35
Anisole	Lead(IV) acetate and boron trifluoride etherate in dichloromethane	2,2′- (1%), 2,4′- (6%), and 4,4′-dimethoxybiphenyl (30%)	23, 36
Anisole	Lead(IV) acetate and boron trifluoride	2,2′- (0.5%), 2,4′- (3%), and 4,4′-dimethoxybiphenyl (2%)	36
2-Methoxytoluene	Electrolysis in aqueous acid	4,4′-Dihydroxy-3,3′-dimethylbiphenyl and its monomethyl ether	37
3-Methoxytoluene	Electrolysis in aqueous acid	4-Hydroxy-4′-methoxy-2,2′-dimethylbiphenyl (19%)	37
4-Methoxytoluene	Electrolysis in aqueous acid	2,2′-Dimethoxy-5,5′-dimethylbiphenyl and 2-hydroxy-2′-methoxy-5,5′-dimethylbiphenyl (34%)	37
4-Methoxytoluene	Manganese(IV) oxide and sulfuric acid	2,2′-Dimethoxy-5,5′-dimethylbiphenyl	38
1,2-Dimethoxybenzene	Electrolysis in aqueous acid	3,3′,4,4′-Tetramethoxybiphenyl and a hydroxytrimethoxybiphenyl (low yield)	39
1,2-Dimethoxybenzene	Aluminum chloride and acetyl chloride in toluene	2,3,6,7,10,11-Hexamethoxytriphenylene (0.08%)	40
1,2-Dimethoxybenzene	Moist iron(III) chloride	2,3,6,7,10,11-Hexamethoxytriphenylene (7%)	41
1,2-Dimethoxybenzene	Chloranil (and related quinones) in 70% v/v aqueous sulfuric acid	2,3,6,7,10,11-Hexamethoxytriphenylene (73%)	42
1,2-Dimethoxybenzene and 3,3′,4,4′-tetramethoxybiphenyl	Chloranil in 70% v/v aqueous sulfuric acid	2,3,6,7,10,11-Hexamethoxytriphenylene (154% based on 1,2-dimethoxybenzene alone)	42
1,3-Dimethoxybenzene	Aluminum chloride in nitrobenzene	2,2′,4,4′-Tetramethoxybiphenyl (2.5%)	43
1,4-Dimethoxybenzene	Diisopropyl peroxodicarbonate and aluminum chloride	2,2′,5,5′-Tetramethoxybiphenyl (30%)	44
2,5-Dimethoxytoluene	Electrolysis in aqueous acid or chromic acid	2,2′,5,5′-Tetramethoxy-4,4′-dimethylbiphenyl	37, 45

166

1,2,4-Trimethoxybenzene	Acenaphthenequinone or phenanthrene-9,10-quinone in polyphosphoric acid	2,2',4,4',5,5'-Hexamethoxybiphenyl (60%)	46
1,2,4-Trimethoxybenzene	Electrolysis in aqueous acid	2,2',4,4',5,5'-Hexamethoxybiphenyl (85%)	47
1,2,4-Trimethoxybenzene	"Almost any oxidizing agent" in acid solution, e.g., chromic acid, iron(III) chloride, iodine chloride	2,2',4,4',5,5'-Hexamethoxybiphenyl	47
1-Methoxynaphthalene	Lead(IV) acetate	4,4'-Dimethoxy-1,1'-binaphthyl	48
1-Methoxynaphthalene	Aluminum chloride in nitrobenzene	4,4'-Dimethoxy-1,1'-binaphthyl (32%)	43, 49
1-Methoxynaphthalene	Concentrated sulfuric acid, or mixed nitric, sulfuric, and acetic acids	4,4'-Dimethoxy-1,1'-binaphthyl	50
1-Methoxynaphthalene	Peroxoformic acid	4,4'-Dimethoxy-1,1'-binaphthyl (70%)	51
1-Methoxynaphthalene	Manganese(III) acetate	1,1'-Diacetoxy-4,4'-dimethoxy-2,2'-binaphthyl	25
1-Methoxynaphthalene	Nitrogen dioxide in carbon tetrachloride	4,4'-Dimethoxy-3,3'-dinitro-1,1'-binaphthyl (28%)	52
1-Ethoxynaphthalene	Benzenesulfonic acid in nitrobenzene	4,4'-Diethoxy-1,1'-binaphthyl (44%)	53
1-Ethoxynaphthalene	Aluminum chloride in nitrobenzene	4,4'-Diethoxy-1,1'-binaphthyl (70%)	43, 54
1-Ethoxynaphthalene	Cation radical from 4,4'-diethoxy-1,1'-binaphthyl	4,4'-Diethoxy-1,1'-binaphthyl	19
1-Ethoxynaphthalene	Potassium anthraquinone-1,5-disulfonate, boric acid, and concentrated sulfuric acid	4,4'-Diethoxy-1,1'-binaphthyl (84%)	55
1-Ethoxynaphthalene	Isatin in concentrated sulfuric acid	4,4'-Diethoxy-1,1'-binaphthyl (11%)	56
1-Ethoxynaphthalene	Isatin in 84% sulfuric acid	4,4'-Diethoxy-1,1'-binaphthyl (22%)	57, 58
1-Ethoxynaphthalene	Phenanthrene-9,10-quinone in sulfuric acid	4,4'-Diethoxy-1,1'-binaphthyl (65%)	57
2-Methoxynaphthalene	Silver fluoride and iodine	2,2'-Dimethoxy-1,1'-binaphthyl (30%)	59
2-Methoxynaphthalene	Aluminum chloride in nitrobenzene	2,2'-Dimethoxy-1,1'-binaphthyl (5%)	43

11.4. FEEDBACK

In literature searching, as in all information operations, it is most important for the abstractor to get feedback from his users. For example, did the search meet the need, and, if so, can the client place a realistic dollar value on the true value of the search to him? Does the client believe that time was saved or needless duplication of effort avoided? Was anything of importance omitted? Was anything of an extraneous nature included? What percentage of the material included was relevant? Should the search be continued on a current basis?

A questionnaire should go out along with the search report to the client so that the abstractor can obtain the necessary feedback. The client should be asked to wait at least a week or two before responding so that he can accurately evaluate the report and its impact.

chapter 12
Auxiliary Services

Some abstracting services, especially the larger published ones, offer a variety of auxiliary services. Some are free to subscribers; others are offered on a fee basis. To the abstractor, they offer exceptionally interesting opportunities for career diversification, challenge, and experimentation.

To the editor or manager of the service, they offer opportunities to provide clients with better services, often of a custom- or tailor-made nature. It is important, however, for the editor or manager to avoid over-extending his resources. He must also be sure that the auxiliary services (unless strictly experimental) will have some degree of continuity. To be successful, they cannot be offered on an on-again, off-again basis.

To the client or customer, the benefits can be substantial, but, of course, he usually must pay substantial additional fees for the services desired.

Some of the auxiliary services offered are:

1. Special bibliographies (often published as part of a regular series, for example, by the American Society for Metals).

2. Photocopy or microcopy service for entire current input, for selected parts (e.g., all patents), or for individual documents. (For example, photocopies of certain original Russian scientific papers can be obtained from Chemical Abstracts Service.)

3. Microfilms of the abstracts. (This is a very popular service since, when coupled to automated microfilm reader-printers, it offers rapid access to abstracts.)

4. Consultation on services supplied. Also, training courses on preparation and use of indexes.

5.   Special abstracting and indexing on a charge basis.

6.   Thesaurus (dictionary of terms used in indexing) and other guides to index use (e.g., the *Index Guide* of *Chemical Abstracts* and *Thesaurus of Metallurgical Terms* of the American Society for Metals).

7.   Indexes (These are sometimes sold separately from the abstracting service proper)—weekly, monthly, semiannual, annual, cumulative (e.g., over a five- or ten-year period).

8.   Current announcement services (daily, weekly, or monthly). This is offered in various forms, for example, a title announcement service like *Chemical Titles,* a publication of Chemical Abstracts Service, or abstracts on 3×5 cards like the Engineering Index Card-A-Lert card service.

9.   Lists showing which libraries subscribe to the journals in the field covered by the service (e.g., *Access* issued by Chemical Abstracts Service).

10.   Translation service.

11.   Subsections (e.g., the Organic Chemistry Section Grouping of *Chemical Abstracts*).

12.   Repackaging of parts of the full service (or issuing and supplementing in special ways) intended for use by groups with special needs and interests. For example, BIOSIS issues *Abstracts of Mycology* and *Abstracts of Entomology,* which are comprehensive, discipline-oriented services designed to meet the needs of specialists whose interests lie within the scope of a larger scientific field. Other examples include:

American Petroleum Institute
   Abstracts of Transportation Storage Literature and Patents
   Abstracts of Petroleum Substitutes Literature and Patents
   Abstracts of Air and Water Conservation Literature and Patents

American Society for Metals
   World Aluminum Abstracts

Chemical Abstracts Service
   Plastics Industry Notes
   Polymer Science and Technology: Patent Edition and Journal Edition
   Chemical-Biological Activities

13.   Computer-readable versions, for example, *COMPENDEX* which is the complete *Engineering Index* on magnetic tape. Almost all major abstracting services offer at least part of their output in computer-readable form.

14.   Custom searches of the information store on specific subjects on request (e.g., as offered by BIOSIS).

15.   Selective current awareness notification or dissemination of in-

formation based on interests of specific individuals or groups (e.g., the "ASCA" service of the Institute for Scientific Information, Philadelphia).

Of the auxiliary services noted above, those that are currently the focus of most of the activity and interest involve the use of computer-readable tapes to provide retrospective search capability and especially selective current awareness notification (see Section 14.5).

chapter 13
Bridging the Time Gap
with Abstracts

There is a very substantial time gap between the inception of research and the publication of results in a journal or presentation of these results at a technical meeting. And there is a further delay before published abstracts are readily available from conventional abstracting services. Special kinds of abstract services can help to bridge the time gap.

13.1. RESEARCH RESULTS SERVICE

A means for bridging the time gap between the completion of work and publication is exemplified by the Research Results Service (RRS) of *Industrial and Engineering Chemistry*. Initiated in January, 1962, this is a rapid informal service that makes available brief abstracts of unpublished research papers in the process of being reviewed for *Industrial and Engineering Chemistry* or one of its several affiliated publications. Subscribers may order copies of manuscripts of interest. Similar services have been offered by other publications in fields other than chemistry.

Kuney and Anderson (17) have described the service:

> "The RRS summary . . . tries to give, in 50 words or less, enough information to let the subscriber decide whether or not he wishes to purchase a copy of the unreviewed, unedited manuscript. These Research Results Service summaries are published as promptly as possible; in some cases it has been possible for a subscriber to obtain a manuscript within six weeks after it has reached the editor's desk. . . .

Manuscript copies for RRS customers are photoprints, run off individually as orders are received. . . .

"Most of the summaries published in the Research Results Service of *I&EC* are prepared or 'assembled' by the RRS staff. Finding the necessary parts somewhere in the manuscript (very often in the conclusion) is accomplished without much difficulty as a rule, and the resulting summary (usually in the author's own words) gives the prospective customer a pretty good idea of what will be in the package he orders. The number of text pages, figures, and tables are included, so that the reader has an even better idea of the type of material he can expect.

"All summaries are submitted to the authors for approval before they are published. As authors become familiar with the Research Results Service type of summary, more summaries are being submitted in acceptable form by authors.

"At the time that an author is asked to approve publication of the Research Results Service summary, he is also reminded that even if his paper is not accepted for publication it can be made available through RRS for 'at least 90 days' after the editor's decision if he is willing. Authors usually agree to such an arrangement if they approve RRS listing in the beginning—and more than two-thirds of the authors do. However, some authors prefer to say 'no'; if so, no effort is made to persuade them to change their minds. . . .

"Prospective users are cautioned that they are ordering unreviewed and unedited manuscripts, but *I&EC*'s editorial practice includes a built-in safeguard against their receiving material of marginal interest. No manuscript is listed unless, and until, one of the *I&EC* editors decides that it should be critically reviewed. As a rule, he has looked over the manuscript, and in all cases he has seen the abstract. This precaution eliminates almost all manuscripts on subjects outside the scope of the journals or unsuitable for other obvious reasons.

"Not all manuscripts withheld from the RRS columns are absent because the author has not been willing to participate. Sometimes the time interval from receipt to acceptance and publication is too short to make an RRS summary practical. Present policy is not to list an article already scheduled for publication at the time the RRS lists go to press (a month before the issue appears). It is felt that a subscriber should not have to go to the expense of ordering a manuscript that he will find in published form in the next issue of *I&EC*. . . .

"Subscribers who order RRS manuscripts are advised of the conditions of purchase both in a paragraph which heads the monthly RRS section and by a statement which appears on each copy sold. These

All papers listed are being considered for publication in I&EC or one of the I&EC Quarterlies. They are available in manuscript form, with the requirement that any reference to their content in a publication must have the author's prior approval.

Orders are processed within 24 hours of receipt, with shipment by first class mail. Prices for I&EC subscribers and nonsubscribers are listed with each manuscript. *Please include payment with order.*

An Improved Equation of State. A new modification of the equation of Redlich and Kwong adequately represents the critical isotherm and is applicable to substances with large acentric factors. Computer programs for pure substances and mixtures are presented in Report UCRL-19011, Lawrence Radiation Laboratory, Berkeley.

Otto Redlich, Victoria B. T. Ngo, University of California and Lawrence Radiation Laboratory, Berkeley

Ms. 69-356   14 pages (3 tables)
I&EC Subscribers $4.00   Nonsubscribers $8.00

Vapor Phase Oxidation as a Process for Raising Octane Number. The process has been applied to five different virgin naphthas, a hydroformate, and a catalytic naphtha. The products, containing as much as 46 volume % of oxygenated compounds, on dehydration give a more orthodox fuel.

Jennings H. Jones, Merrell R. Renske, and Russell A. Rusk, The Pennsylvania State University

Ms. 69-508   25 pages (2 figures, 11 tables)
I&EC Subscribers $6.00   Nonsubscribers $12.00

Anhydrous Aluminum Bromide in Organic Solvents. Molten material was added to cyclohexane without any hazardous conditions being observed. The loss of $AlBr_3$ value in solution was checked for a limited time. The effects of light, moisture, and heat on the stability of the solution were investigated.

Punkajkumar M. Trivedi, Michigan Chemical Corp.

Ms. 69-559   19 pages (2 figures, 2 tables)
I&EC Subscribers $4.00   Nonsubscribers $8.00

Radical Polymerization of Allylisocyanurates with Vinyl Monomers. Radical copolymerization of di- or tris-(N-2-hydroxy,3-allyloxypropyl)-isocyanurates (DHAIC) and (THAIC) with methyl methacrylate and styrene was studied, and rate constants were evaluated. Methyl methacrylate copolymerizes with DHAIC and THAIC at a substantially higher rate than styrene.

H. Alaminov, M. Michailov, and T. Michneva, Sofia, Bulgaria

Ms. 69-560   10 pages (3 figures, 2 tables)
I&EC Subscribers $2.00   Nonsubscribers $4.00

Influence of Surface Turbulence and Surfactants on Gas Transport Through Liquid Interfaces. Results using a new technique indicate that at high turbulence rates the statistical nature of interfaces may be described by a Danckwerts-type distribution function of surface ages. Nature of surface films and their stability against interfacial turbulence is discussed.

Thomas G. Springer and Robert L. Pigford, University of California, Berkeley

Ms. 69-562   46 pages (10 figures, 5 tables)
I&EC Subscribers $10.00   Nonsubscribers $20.00

Hydrogen Reduction of Alkali Sulfate. The hydrogen reduction of sulfate dissolved in molten alkali carbonate (600° to 850°C) is autocatalyzed by sulfide and catalyzed by iron compounds. A kinetic study, which showed the reduction rate to be independent of sulfate concentration, suggests that polysulfide is an intermediate.

J. R. Birk, C. M. Larsen, W. G. Vaux, and R. D. Oldenkamp, Atomics International North American Rockwell Corp.

Ms. 69-565   26 pages (8 figures, 1 table)
I&EC Subscribers $6.00   Nonsubscribers $12.00

Intraparticle Diffusion and Nonlinear Kinetics in Fixed Beds. Presents an algorithm which seems well suited to use with digital simulation languages (such as IBM 360 CSMP). Intraparticle diffusion, convective mass transfer between particle and fluid, and homogeneous nonlinear chemical reaction within the particle are accounted for.

James M. Wheeler, Eastman Kodak Co., and Stanley Middleman, University of Massachusetts

Ms. 69-566   20 pages (4 figures)
I&EC Subscribers $4.00   Nonsubscribers $8.00

Study of Interaction of Solvent and Catalyst in Coal Hydrogenation. Catalyst impregnated subbituminous Wyoming coal was hydrogenated in the presence of different solvents. Catalysts markedly improved conversions only with poor solvents. The solvent which had a nonpolar solubility parameter of approximately 9.5 appeared most effective. Coal size didn't affect this conclusion.

G. R. Pastor, J. M. Angelovich, and H. F. Silver, University of Wyoming

Ms. 69-567   9 pages (1 figure, 1 table)
I&EC Subscribers $2.00   Nonsubscribers $4.00

Macromolecular Ultrafiltration with Microporous Membranes. Discusses factors controlling membrane flux and macromolecular rejection efficiency. Flux is limited by mass transfer conditions on feed solution side of membrane. Macromolecular rejection by membrane is determined by geometric properties of membrane pores and macromolecules in solution. Presents ultrafiltration data for macromolecular solutions.

Robert L. Goldsmith, Abcor, Inc.

Ms. 69-576   57 pages (20 figures)
I&EC Subscribers $12.00   Nonsubscribers $24.00

Fluid Mixing Characteristics in Void Spaces in Packed Beds. Local and time variations of concentration were observed at the exit of two-dimensional packed bed. Local scale of per-

turbulence was about 0.29 Dp, and time-scale of variation showed the origin of turbulence might be due to eddy shedding.

Motoyuki Suzuki and Daizo Kunii, University of Tokyo, Japan

Ms. 69-573   20 pages (8 figures)
I&EC Subscribers $4.00   Nonsubscribers $8.00

Characterization of Liquid Solid Reactions—Hydrochloric Acid–Calcium Carbonate Reaction. Reaction rate for flow through a reactive channel is predicted for first and second order reversible and irreversible heterogeneous reactions. An approximate boundary layer solution comparable to the exact solution is developed.

B. B. Williams, Esso Production Research Co., J. L. Gidley, Humble Oil & Refining Co., J. A. Gunn, Purdue University, and R. S. Schechter, University of Texas

Ms. 69-577   28 pages (6 figures, 5 tables)
I&EC Subscribers $6.00   Nonsubscribers $12.00

Comparison of Moments, s-Plane, and Frequency Response Methods for Analyzing Pulse Testing Data from Flow Systems. The three methods are used to fit the gamma distribution model with bypassing to experimental data obtained in a flow channel agitated by air bubbles. The s-plane analysis is recommended as the best of the three methods.

J. L. Johnson, L.T. Fan, and Y.S. Wu, Kansas State University

Ms. 69-578   61 pages (4 figures, 5 tables)
I&EC Subscribers $14.00   Nonsubscribers $28.00

Carbon Black Feedstock from Low Temperature Carbonization Tar. Selective hydrotreatment removes oxygen, nitrogen, and sulfur from LTC tar to produce a full-range, highly aromatic material from which carbon black feedstock is

recovered by fractionation. Preliminary economics based on pilot data are presented.

Donald C. Berkebile, Harold N. Hicks, and W. Sidney Green, Ashland Oil and Refining Co.

Ms. 69-579   29 pages (3 figures, 8 tables)
I&EC Subscribers $6.00   Nonsubscribers $12.00

Quasisteady-State Stagnant Film Evaporation with Moving Interface. A general model was developed for an evaporating liquid. Open and closed systems with variable transport area were considered. The model agreed with results from a slowly evaporating closed system of constant transport area.

James M. Pommersheim, Bucknell University

Ms. 69-580   22 pages (2 figures)
I&EC Subscribers $6.00   Nonsubscribers $12.00

Quasilinearization and Estimation of Parameters in Partial Differential Equations. Lee's technique is extended to partial differential equations with nonlinear boundary conditions. Convective and radiative heat transfer coefficients are obtained from an equation that models the transient heating of a bed of spheres.

V. G. Fox, L. W. Ross, and K. D. Van Zanten, University of Denver

Ms. 69-583   8 pages (1 figure, 1 table)
I&EC Subscribers $2.00   Nonsubscribers $4.00

Mass Transport in Reverse Osmosis in Case of Variable Diffusivity. Effects of practical ranges of linear, parabolic, and exponential diffusivity-concentration relationships on the concentration polarization in laminar flow were investigated. Theoretical results are interpreted in terms of their design application.

Y. T. Shah, University of Pittsburgh

Ms. 69-584   20 pages (7 figures)
I&EC Subscribers $4.00   Nonsubscribers $8.00

Fig. 13.1. Use of indicative abstracts to help make information quickly available. *Source. Industrial and Engineering Chemistry*, **62**, February 1970, p. 67. Published by The American Chemical Society, Washington. Reprinted with permission of copyright owner.

175

conditions are: (1) the subscriber will treat the manuscript as a personal or private communication; and (2) reference to its content will not be made in any publication without the author's prior approval. No problems have been encountered so far with the manuscripts listed. . . ."

In the period since Kuney and Anderson wrote their article, the procedures used by the RRS staff have changed somewhat. Present practice is that the authors of papers prepare the abstracts themselves using a "do-it-yourself" form; these abstracts are usually reasonably satisfactory. Also, the maximum number of words allowed in the abstract has been cut from 50 to 40 words. It has been found that this is adequate inasmuch as the title is now part of the summary, which was not the case in the first year or so. Another procedural change is that the author now signs a form stating that the manuscript can be made available in manuscript form through RRS until published or withdrawn by the author.

Figure 13.1 illustrates the kinds of abstracts used (indicative) and the method of presentation.

Beginning in 1971 RRS is scheduled to be offered as part of the new "Single Article Announcement Service" of the American Chemical Society.

13.2. CURRENT ALERTING ABSTRACTING SERVICES

Services such as *Polymer Science and Technology, Chemical-Biological Activities, Current Abstracts of Chemistry and Index Chemicus,* and the *Derwent Central Patent Index* are current alerting abstracting services. The editors of such services make every effort to supply information to the user as quickly as possible. Computer-based sorting and printing of abstracts and indexes speed up the processes in most cases. In addition, special efforts are made to obtain advance copies of documents either as galleys or page proof and via air mail. These services will go to such lengths as the hand copying of material to be abstracted; for example, Derwent personnel do this if there is no other way to obtain the material found in patents. The abstracts are often kept relatively brief (up to 100 words) since this makes for faster writing, typing, printing, and issuance of the bulletin. Detailed abstracts, if needed, can come later in more conventional, but slower kinds of services.

13.2.1. Current Abstracts of Chemistry and Index Chemicus

A good example of express service operations is *Current Abstracts of Chemistry and Index Chemicus (CAC),* a publication of the Institute for

PRESENTATION OF ABSTRACTS

A typical abstract is reproduced below. Each abstract gives the article's title [A], the names of its authors [B], their addresses [C], the journal citation [D], the date the article was received for publication [E], and the language of the full text if other than English [F]. The author's narrative summary [G] is supplemented by all useful diagrams [H] from the full text. Analytic techniques [I] and potential and proven uses [J] of the compounds are indicated by appropriate data symbols. A star [K] indicates abstracts of articles reporting the synthesis or use of labeled compounds. Circled arabic numerals [L] are used to identify the structural diagrams of new compounds. Circled numerals are joined by a dash [M], as ④ — ⑥ (meaning ④, ⑤, and ⑥, to indicate that the diagram gives rise to more than one specific compound. Although molecular formulas of all specific compounds represented by the diagrams do not appear with the abstracts, the circled numerals will be used as suffixes to abstract numbers (as 123456-4) in the molecular formula indexes of *Index Chemicus*, and in the magnetic tape and printed Wiswesser Line Notation indexes of the *Index Chemicus Registry System®*, to allow retrieval of information on specific compounds. A circled asterisk ✱ indicates formulas of old compounds produced by a new reaction or a new synthesis.

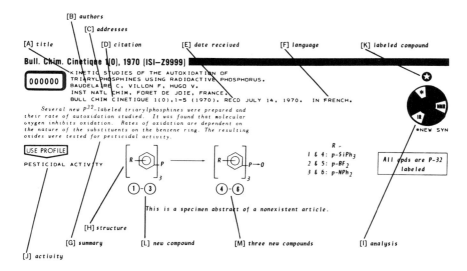

Fig. 13.2. Examples of "express" abstracts. *Source. Current Abstracts of Chemistry and Index Chemicus,* **36**, No. 1, Issue No. 319, January 7, 1970. Published by Institute for Scientific Information, Philadelphia. Reprinted with permission.

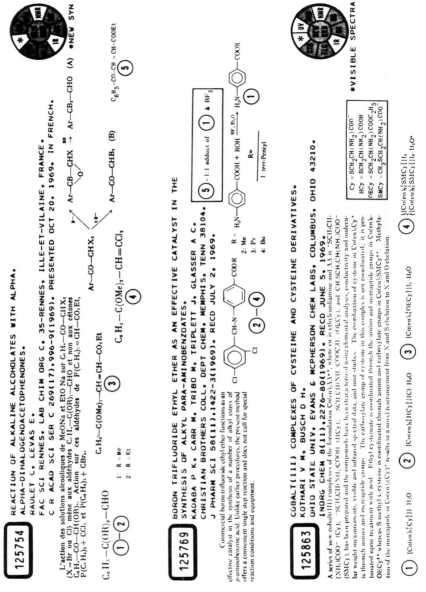

125754

REACTION OF ALKALINE ALCOHOLATES WITH ALPHA, ALPHA-DIHALOGENOACETOPHENONES.

RAULET C., LEVAS E.

FAC SCI RENNES, LAB CHIM ORG C, 35-RENNES, ILLE-ET-VILAINE, FRANCE.

C R ACAD SCI SER C 269(17),996-9(1969). PRESENTED OCT 20, 1969. IN FRENCH.

L'action des solutions alcooliques de MeONa et EtO Na sur C₆H₅—CO—CHX₂ (X=Br ou Cl) mène aux aldéhydes C₆H₅—C(OR)₂—CHO et non aux cétones C₆H₅—CO—CH(OR)₂. Action sur ces aldéhydes de P(C₆H₅)₃=CH—CO₂Et, P(C₆H₅)₃+ CCl₄ et P(C₆H₅)₃+ CBr₄.

C₆H₅—C(OR)₂—CHO

①—②

1: R = Me
2: R = Et

C₆H₅—C(OMe)₂—CH—CO₂Et

③

C₆H₅—C(OMe)₂—CH=CH—CO₂Et

④

Ar—CB—CHX → Ar—CB₁—CHO (A) *NEW SYN

B→\ /←B'
   O

Ar—CO—CHX₂

Ar—CO—CHB, (B)

C₆H₅—CO—CH = CH—COOEt

⑤

* IR NMR

125769

BORON TRIFLUORIDE ETHYL ETHER AS AN EFFECTIVE CATALYST IN THE SYNTHESIS OF ALKYL PARA-AMINOBENZOATES.

KADABA P K, CARR M, TRIBO M, TRIPLETT J, GLASSER A C.

CHRISTIAN BROTHERS COLL, DEPT CHEM, MEMPHIS, TENN 38104.

J PHARM SCI 58(11),1422-3(1969). RECD JULY 2, 1969.

Commercial boron trifluoride ethyl ether functions as an effective catalyst in the synthesis of a number of alkyl esters of p-aminobenzoic acid. Unlike earlier procedures, the present method offers a convenient single step reaction and does not call for special reaction conditions and equipment.

COOR

H₂N—◯—COOH + ROH $\xrightarrow{BF_3 \cdot Et_2O}$ H₂N—◯—COOR

R =
2: Me
3: Pr
4: Bu

⑤ 1:1 adduct of ① & BF₃

R=
1 tert-Pentyl

125863

COBALT(III) COMPLEXES OF CYSTEINE AND CYSTEINE DERIVATIVES.

KOTHARI V M, BUSCH D H.

OHIO STATE UNIV, EVANS & MCPHERSON CHEM LABS, COLUMBUS, OHIO 43210.

INORG CHEM 8(11),2276-80(1969). RECD JUNE 5, 1969.

A series of new cobalt(III) complexes of the formulation Co(en)₂AA⁺⁺, where en is ethylenediamine and AA is ⁻SCH₂CH(NH₂)COO⁻ (Cy), ⁻SCH₂CH₂NH₂COOH (HCy), ⁻SCH₂CH(NH₂)COOC₂H₅ (OECy), and CH₃SCH₂CH(NH₂)COO⁻ (SMCy), has been prepared and the compounds have been characterized using elemental analyses, conductivity and molecular weight measurements, visible and infrared spectral data, and nmr studies. The coordination of cysteine in Co(en)₂Cy⁺ is through amino and mercaptide groups. The carboxylate group of cysteine in this complex is not coordinated; it is protonated upon treatment with acid. Ethyl cysteinate is coordinated through the amino and mercaptide groups in Co(en)₂OECy⁺⁺ whereas S methyl cysteine is coordinated through amino and carboxylate groups in Co(en)₂SMCy⁺⁺. Methylation of the mercaptide in Co(en)₂Cy⁺ results in a novel rearrangement from N and S chelation to N and O chelation.

Cy = SCH₂CH(NH₂)COO⁻
HCy = SCH₂CH(NH₂)COOH
OECy = SCH₂CH(NH₂)COOC₂H₅
SMCy = CH₃SCH₂CH(NH₂)COO⁻

① [Co(en)₂HCy]Cl₂·H₂O ② [Co(en)₂HCy]Cl·H₂O ③ [Co(en)₂OECy]I₃·H₂O ④ { [Co(en)₂SMCy]I₃ }, { [Co(en)₂SMCy]I₃·H₂O }

* UV NMR IR *VISIBLE SPECTRA

Fig. 13.2. *(Continued)*

178

Scientific Information in Philadelphia. This service makes effective use of author abstracts (instead of rewriting these). All titles appear in English. But author-prepared abstracts (if available in the journal) appear in the language published without any further translation or editing. The use of untranslated foreign language abstracts is not regarded as a barrier to the reader because of the use of chemical structure diagrams with entries. These diagrams are usually not redrawn but instead are clipped out of the original whenever available in satisfactory form; they are a kind of universal language to chemists and convey a substantial amount of information. For new compounds, if the author does not supply diagrams, the structures will be drawn.

In the past, abstract proofs were sent to authors by the editorial staff for approval. This helped to ensure accuracy and made the abstracts in some instances more accurate than the original article, where a mistake may have been overlooked. In order to save time, this step is now omitted.

CAC uses unique graphic displays such as their "Use Profile," and "Instrumental Data Alert." The "Use Profile" was conceived to alert the reader to biological test data as, for example, "Antibiotic Activity." The "Instrumental Data Alert" indicates which types of instrumental analyses were used and alerts the reader to new chemical reactions reported. (See Fig. 13.2.)

Machine-readable input for *CAC* includes the citation, molecular formulas, index terms, "Use Profile," and "Instrumental Data Alert." Since the material is machine readable, this makes for ready manipulation, such as rapid cumulation of indexes and automatic searching capability. See Fig. 13.3 for an example of the kind of indexing used in *CAC*; these indexes appear monthly and are cumulated semiannually and annually.

Structural diagrams are available in machine-readable form through a service known as the *Index Chemicus Registry System (ICRS),* which, like *CAC*, is a product of the Institute for Scientific Information. This service provides all structural diagrams in encoded form and is distributed to subscribers monthly both on magnetic tape (which also contains the machine-readable input from *CAC*) and as a printed index.

Another way of achieving speed is by limiting the coverage of a service to clearly delineated and specific fields or groups of journals. For *CAC*, about 100 key journals in organic, pharmaceutical, medicinal and biological chemistry are abstracted for all new compounds and reactions reported therein. This kind of careful selectivity is essential to achieving both speed and quality.

As a result of efforts such as those described, abstracts are said to appear about 30 days after a journal is received at *CAC*.

PERMUTERM INDEX

PYROLYSIS - - - - 125959

ACETALDEHYDE
ALDOL RXN - - - - 127164
CYCLOPROPANE - - "

ACETALDIMINE
ACETAMIDE - - - - 127823
ACETYL - - - - - "
TRICHLORO - - - - "
TRICHLOROETHYL - "

ACETAMIDE
ACETALDIMINE - - 127823
ACETYL - - - - - "
ACID HYDROLYSIS - 127018
HYDANTOIN - - - - 125953
PHOSPHONOACETATE 125954
POLYMERIZATION - 125953
ROTATIONAL ISOMER 126330
SYNTHESIS - - - - 125953
TRICHLORO - - - 127823
TRICHLOROETHYL - "
WITTIG-HORNER RXN 125954

ACETAMIDINE
BETAINE - - - - - 127650
CHLOROSULFONYL - "

ACETANILIDE
ISATIN - - - - - 125931

ACETATE
ACRYLATE - - - - 127803
ALCOHOL - - - - - 127320
AMINO ACID - - - 126268
CHLORO - - - - - 127727
COBALT - - - - - 126073
COMPLEX - - - - - 127727
COPPER - - - - - "
COPPER COMPLEX - 126268
CYCLOBUTANE - - 126530
FUMARATE - - - - 127803
GLUTAMIC ACID - - 128189
LEWIS ACID - - - 127320
NICKEL - - - - - 126073
POLYMER - - - - - 127803
PYRIDINE - - - - 127727
SOLVOLYSIS - - - 126530
STYRENE - - - - - 127803
SUBSTITUTED - - 127727
TITANIUM COMPLEX 127320
TOSYLATE - - - - 126530

ACETIC ACID
ACETIMIDE - - - - 126621
ACID HYDROLYSIS - 127017
ALKYL ESTER - - - 127290
ALKYLATION - - - 127835
AMINOMETHYL - - - 125970
AZABICYCLONONANE "
BENZENE - - - - - 126272
BENZONORCARADIENE 127896
BICYCLIC - - - - 127895
BINARY EQUILIBRIA 127374
BUTANOIC ACID - - 126649
BUTYRIC ACID - - 128262
CATALYSIS - - - - 128344
CHLORIDE - - - - "
CHLORINE - - - - 126776
CHROMIUM COMPLEX 125903
COMPLEX - - - - - 127859
CYCLOHEXANOIC ACI 126649
CYCLOPENTANE - - 125970
DIAZOKETONE - - - 127896
DIETHOXYMETHANE - 127017
DIFLUORO - - - - 127865
ESTER - - - - - - 126272
127017
127865
ESTERIFICATION - 127157
ETHANESULFONIC - 127859
ETHYL ACETATE - - 126183

1

PICRIC ACID - - - 126456
POLYMER - - - - - 126570
PROPENE - - - - - 125787
PROPIONAMIDE - - 125912
THERMAL - - - - - 128240
THIOCYANATE - - - 125914
TRIAZINE - - - - 125913
125914
TRIFLUORO - - - - 128240
VANADIUM COMPLEX 127961

ACETOPHENONE
ACETONITRILE - - 125912
ALCOHOLATE - - - 125754
ALDEHYDE - - - - 126348
BENZALDEHYDE - - 125912
126300
BENZOPHENONE - - 125912
BIPHENYL - - - - 126085
CARBONYL - - - - 126300
CARBONYL CPD - - 126824
CHYMOTRYPSIN - - 127311
CONDENSATION - - 126348
COPPER COMPLEX - 126587
CRYSTAL STRUCTURE "
EFFECT - - - - - 126167
GRIGNARD RXN - - 126824
HALIDE - - - - - "
HALO - - - - - - 125977
IONIZATION - - - 126167
MAGNESIUM - - - - 126824
NEW SYN - - - - - 125754
PHENYLACYL HALIDE 127311
PHOSPHAZO - - - - 127825
PHOSPHORESCENCE - 126300
PROPIONAMIDE - - 125912
QUINOLONE - - - - 125916
SUBSTITUENT - - - 126167
SULFONE - - - - - 125974
SULFONIUM YLIDE - 127311
TRIPHENYL - - - - 127825
UV - - - - - - - 126085

ACETOXIME
PLATINUM COMPLEX 127981

ACETOXONIUM CPD
CYCLOPENTANE - - 127720
FLUOROBORATE - - "

ACETOXY
BENZO - - - - - - 127855
BENZOBARRELENE - "
BICYCLOOCTADIENE "
BROMINATION - - - "
BROMO - - - - - - "
TETRAFLUORO - - - "

ACETOYLSIS
CELLULOSE - - - - 126512
POLYMERIZATION - "

ACETYL
ACETALDIMINE - - 127823
ACETAMIDE - - - "
ACTIVITY - - - - 126171
ANTIBACTERIAL - - "
ANTIBIOTIC - - - "
DERIVATIVE - - - "
DICHLORO - - - - 127831
PHOSPHONIC ACID - "
PHOSPHONYL - - - "
PHOSPHORIC ACID - "
SIOMYCIN A - - - 126171
TRICHLORO - - - - 127823
TRICHLOROETHYL - "
VINYL - - - - - - 127831

ACETYLACETONATE
ALLENE - - - - - 127223
AMINE - - - - - - 126584
ANILINE - - - - - "
COBALT COMPLEX - 126230

2

SOLUBILITY - - - 128039
SULFONE - - - - - 125974
TOSYL - - - - - - 126827
TRANSITION METAL 125940
TRIDECATETRAENE - 126632

ACETYLIDE
ETHANE - - - - - 126319
OLIGOMER - - - - "
PROPANE - - - - - "

ACETYLIDE COMPLEX
CARBONYL - - - - 127703
IRON COMPLEX - - "
NICKEL COMPLEX - "

ACETYLTRANSFERASE
CARNITINE - - - - 127429
ORD - - - - - - - "

ACID
ANHYDRIDE - - - - 126128
AZIRANE - - - - - 127833
BENZHYDROXAMIC - 127692
CADMIUM COMPLEX - 125672
CONVERSION - - - 126128
CRYSTAL STRUCTURE 126590
DECOMPOSITION - - 127827
DIAMIDE - - - - - 127838
DISSOCIATION - - 126435
ESTER - - - - - - 127827
127832
127834
127838
HYDROLYSIS - - - 127834
KETONE - - - - - 127832
MIXTURE - - - - - 126435
NAPHTHOLSULFONIC 125672
PERPHOSPHORIC - - 127827
PHOSPHORAMIDOUS - 127834
PHOSPHORAMIDYL - 127833
PHOSPHORDIAMIC - "
PHOSPHORDIAMIDOUS 127834
POLYMER - - - - - 126128
PYRANDICARBOXYLIC 126590
REACTION - - - - 127832
REARRANGEMENT - - 127692
SELENOPHOSPHORIC 127838
SOLID STATE - - - 126128
THIOPHOSPHINIC - 127832
THIOPHOSPHINOUS - "
ZINC COMPLEX - - 125672

ACID A
LABELED CPD - - - 126150
PHOSPHOLIPASE - - "

ACID CHLORIDE
CHLOROMETHYL - - 127824
DIAZOMETHANE - - 126763
METHYL - - - - - 127824
PHOSPHONIC ACID - "
PYRAZOLINE - - - 126763
REACTION - - - - "
SYNTHESIS - - - - "

ACID DISSOCIATION
ABSORPTION SPECTR 127022
PYROMELLITIC ACID "

ACID FLUORIDE
AMIDE - - - - - - 127345
PHOSPHORUS ACID - "

ACID HYDROLYSIS
ACETAMIDE - - - - 127018
ACETIC ACID - - - 127017
DIETHOXYMETHANE - "
ESTER - - - - - - "

ACID PROTEASE
CYCLOPIUM - - - - 128195
PENICILLIUM - - - "

3

IRIDIUM COMPLEX - 127323
KETONE - - - - - 126520
LIGHT SCATTERING 126910
LINCOLNENSIS - - 125715
MALEIC ANHYDRIDE 127323
METABOLITE - - - 125715
METHYLENE - - - - 127882
METHYLTHIO - - - 125715
NEW RXN - - - - - 125802
OXAPHOSPHAZOLINE 127829
OXIDO - - - - - - "
OXO - - - - - - - "
PHOSPHINE - - - - 127323
PHOSPHINYL - - - 127829
POLYMER - - - - - 126910
PROPIONIC ACID - 127092
PYRROLIDINE - - - 127882
SPIRO CPD - - - - 126520
STREPTOMYCES - - 125715
STYRENE - - - - - 127323
VINYL ESTER - - - 126520

ACRYLONITRILE
COPOLYMER - - - - 128202
COUMARIN - - - - 125771
CYCLOTETRASILOXAN 127806
METHACRYLATE - - "
128202
POLYMER - - - - - 126980
127806
STYRENE - - - - - "
128202
VINYL CHLORIDE - "

ACRYLOYL FLUORIDE
ISOCYANATE - - - 125783
NITROSOMETHANE - "
POLYMER - - - - - "
TRIFLUORO - - - - "

ACTINIDE
EXTRACTION - - - 125944
LANTHANIDE - - - "
PHOSPHINE OXIDE - "

ACTINOMYCETE
ISOBORNEOL - - - 126179
METABOLITE - - - "

ACTIVATED
ALUMINUM - - - - 126245
OXIDE - - - - - - "
POWDER - - - - - "
SILICON - - - - - "
SULFIDE - - - - - "
ZINC - - - - - - "

ACTIVATION
CALCITE - - - - - 127571
CRYOLITE-ALUMINA 126834
DIVALENT CATION - 127479
ELECTROLYTE - - - 126834
LATTICE DISORDER "
YEAST ENOLASE - - 127479

ACTIVATION ANALYS
BARIUM - - - - - 126194
INTERNAL STANDARD "
PHOSPHORUS - - - "

ACTIVATION ENERGY
NITROGEN OXIDE - 126441

ACTIVE CARBON
CATALYSIS - - - - 127994
PREPARATION - - - 127391
SORPTION - - - - 127994

ACTIVE CENTER
ANIONIC POLYMERIZ 126572
POLYMETHACRYLATE "

4

PENICILLANIC ACID 125784
PEPTIDASE - - - - 126160
PEPTIDE - - - - - 125712
126160
PHENOTHIAZINE - - 125772
PHOSPHONIC ACID - 127871
POLYMER - - - - - 127315
POTENTIAL - - - - 127871
PROPANEDIAMINE - 126559
126561
PROPIONAMIDE - - 126555
PROPYLAMINE - - - "
PYRANO - - - - - 125772
PYRIDAZINE - - - 125676
RARE-LIKE - - - - 126561
SEDATIVE - - - - 126562
SIOMYCIN A - - - 126171
126172
SPASMOLYTIC - - - 125632
126555
STREPTOMYCES - - 126175
STRUCTURE - - - - 127964
SUBSPORIN A - - - 125712
SUBSPORIN B - - - "
SUBSPORIN C - - - "
SYNTHESIS - - - - 125765
THIENYL - - - - - 125801
THIOLCARBOXYLIC - 126172
THYROXINE - - - - 126154
TOTAL SYNTHESIS - 126020
TUBERCULOSTATIC - 125772

ACTIVITY COEFFICI
IONIC MELT - - - 126388
PAPER - - - - - - "

ACTOMYOSIN
ELECTRON - - - - 127625
MICROSCOPY - - - "

ACUMINATA
ALKALOID - - - - 126529
CEPHAELINE - - - "
CEPHAELIS - - - - "
LABELED CPD - - - "
SITOSTEROL - - - "

ACYL
ACETYLENE - - - - 125965
ACYLOXY - - - - - 127866
ALKYL - - - - - - 125788
AMIDE - - - - - - 125965
AMIDINE - - - - - 127143
AMINAL - - - - - "
ARYL - - - - - - 125996
BUTADIENE - - - - 127866
BUTENOIC ACID - - 125967
CARBAMIDE - - - - 125965
CARBOETHOXY - - - 125967
CARBOXYLIC ACID - 125996
CHYMOTRYPSIN - - 128256
CYCLIZATION - - - 125967
DIHYDRO - - - - - 125996
EPOXY - - - - - - "
ESTER - - - - - - "
ETHOXY - - - - - 127866
ETHYLENE GLYCOL - "
FURAN - - - - - - 125996
GRIGNARD REAGENT 125965
HEXANOIC ACID - - 125996
HYDROXY - - - - - 125967
IOTSICH REACTION 125965
ISOXAZOLIUM CPD - 127143
KETENIMINE - - - "
MALONAMIDE - - - 125967
METHOXY - - - - - "
ORGANOMETAL - - - 125965
OXIDO - - - - - - 127866
PHENYL - - - - - 125965
PHOSPHITE - - - - 127866
PHOSPHOLENE - - - "
PIPERIDINE - - - 125788
PYRIDINE - - - - "

5

ADDUCT
ALUMINUM CPD - - 127878
AMINE - - - - - - 125951
ANTHRACENE - - - 127118
ANTIMONY - - - - 127680
ARSENIC - - - - - "
BENZOQUINONE - - 125951
BENZYNE - - - - - 127118
BICYCLOHEPTANE - 126548
BORON CPD - - - - 127878
BORON TRIHALIDE - 125704
DIELS-ALDER RXN - 127118
ETHENONAPHTHACENE "
FLUORENE - - - - 128315
GUANOSINE - - - - "
HYDROXY - - - - - "
LABELED CPD - - - "
LITHIUM - - - - - 125829
MOLYBDENUM - - - 125579
NORBORNANE - - - 126548
OLEFIN - - - - - 126655
PHENANTHRENE - - "
PHENYL - - - - - 125829
PHOSPHATE - - - - 126548
PHOSPHINE OXIDE - 125579
PHOTO RXN - - - - 126655
POTASSIUM - - - - 125829
PROTEIN - - - - - 125951
QUINONE - - - - - 126655
RNA - - - - - - - 128315
SESQUIFULVALENE - 127715
TRIMETHYLAMINE - 125704
TRIPTYCENE - - - 127118
TROPYLIUM SALT - 127715
X-RAY - - - - - - 125579

ADENINE
ADENOSINE - - - - 127070
AMINO ACID - - - 127138
127139
AMINOACYL - - - - 127138
127139
ARYLALKYL - - - - 127086
BRANCHED-CHAIN - 125841
CARBOHYDRATE - - "
COBALT - - - - - 126768
127070
COMPLEX - - - - - 126768
127070
COPPER COMPLEX - 125838
CYSTEINE - - - - 127500
DEHYDROGENASE - - 126852
127500
DINUCLEOTIDE - - 127086
FLAVIN - - - - - "
GLYCERALDEHYDE - 127500
HYPOXANTHINE - - 127214
IMIDAZOLE - - - - 125807
ISOALLOXAZINE - - 127086
LABELED CPD - - - 127139
127500
MELANOPHORE - - - 127583
MERCAPTOPURINE - 127070
NEW SYN - - - - - 125807
NICOTINAMIDE - - 126852
127500
NITRILE - - - - - 125807
NUCLEIC ACID - - 127214
NUCLEOSIDE - - - 125838
125841
126643
127138
127139
NUCLEOTIDE - - - 127583
PHOSPHATE - - - - 127214
PURINE - - - - - 127138
127139
RHODOSPIRILLUM - 126852
RIBOFURANOSE - - 125807
RIBONUCLEASE - - 127214
RIBOPYRANOSE - - 126643
RUBRUM - - - - - 126852
SPECTRA - - - - - 127070

6

INDEX CHEMICUS

Fig. 13.3. Example of computer-printed index to a current alerting abstracting service. *Source. Index Chemicus, Monthly Cumulative Index,* **36,** Nos. 319–322, January 1970. Reprinted with permission.

13.2.2.   Some Generalizations

We can make several generalizations about express services. First, to try to cover everything makes it almost impossible to achieve optimum results in current alerting publications even in relatively limited fields. Hence, prudent selection of the number and kinds of documents to be covered is important. The selection *by one person* of articles to be covered and abstracting (if satisfactory author abstracts are not available) and indexing *by the same person* also contribute to speed in alerting-service publication. Elimination of repetitive effort, whether mechanical or intellectual, is a key to achieving speed; clipping out abstracts and other material from the original documents is one relatively simple way to do this. Another way is minimal (preferably single) keyboarding of input. Second, in the abstractor's commendable zeal for issuing an express service promptly, he must make every effort to maintain accuracy in his abstracts. Accuracy always takes precedence over speed.

chapter 14

The Abstractor, Computers, and Information Systems

14.1. INTRODUCTION

The contemporary abstractor views computers and other related machines and information systems as allies rather than threats or competitors.

It behooves the abstractor to keep himself as fully informed as possible about significant developments in computer science and information systems. If he does this, he will be in an excellent position to use any such tools available to him to maximum advantage in some phases of his work.

Before discussing specifically what can be done with computers in abstracting and related functions, some preliminary comments on the limitations and requirements for successful computerization are appropriate.

The existence of a computer center (with first-rate personnel and suitable equipment) in one's organization is merely a prerequisite to actual use of the center. For example, in most industrial firms, the computer center, as one part of the parent organization, serves several departments such as manufacturing, research and development, marketing, and accounting, all of which may, in effect, vie for the services of the center. The abstractor should determine whether the resources of the computer center are available to him at all and, if so, at what level of priority. The computer center must have adequate scheduled time (both people-time and machine-time) available to the abstractor if he is to use it to full advantage.

If an abstractor works with a computer-oriented organization (one which has the necessary equipment, personnel, and funds), he will want to give serious thought to getting his abstracts produced in machine-readable form. This requires that the computer be programmed (instructed) to accept the input and to process it as required. All of this seems easy enough on the surface, but the abstractor should know that programming is demanding work and that the preparation of workable computer programs for information storage and retrieval often requires a substantial amount of time and skill. It is not necessary for the abstractor to learn how to program—this is best left to the professional programmer.

The abstractor should also know that while, in theory, almost anything can be put into machine-readable form, the practical constraints of available equipment may impose limitations—limitations that need to be identified as clearly as possible at the outset.

For example, the abstractor needs to know what the specific computers available to him will accept as input. Thus, certain mathematical expressions, scientific symbols, and the like may be used in the body of the document but not in the abstract because they may not be "machineable" with the equipment at hand. In this case, "verbalized" alternatives may have to be used by either the abstractor or the keypuncher, depending on who is given the assignment of effecting the "conversion." Some suggestions are listed below:

Expression		Verbalized Alternative
$\sqrt{a-b}$	use	square root of $(a-b)$
cm^2	use	sq cm
$\rightarrow$	use	yields
$32°$	use	32 deg.

Cost is another factor that needs investigation. Will the abstractor incur an internal accounting charge against his department when he uses computer center resources? (The answer is probably yes.) If so, at what rates? Cost estimates need to be worked out ahead of time to insure that the budget is adequate for the jobs to be done. Since initial investments in computer center time can be very expensive, a decision to computerize an operation must be made with great care. If it is concluded, on the basis of a feasibility study, that computer use will "pay for itself" within a reasonable period of time or that noncomputer procedures cannot continue to do the job adequately, the decision is relatively easy.

Since almost all of the major external published abstracting services have computerized some of their operations, personal visits to one or more of these services is advisable before launching in-house computerization.

There are a number of functions related to abstracting in which computers and other related machines and systems either already play a role at present or are likely to play an increasing role in the future. Some of these are discussed in the paragraphs that follow.

14.2. COMPUTER-PRINTED INDEXES TO ABSTRACTS

Without question, the principal impact of computers on abstracting so far has been in speeding up the printing of indexes, especially cumulative indexes to abstracts. Machines, when properly instructed by humans, can sort, cumulate, and print rapidly and efficiently. This means that indexes, especially cumulative indexes, can appear both more rapidly and at more frequent intervals (e.g., monthly instead of annually) and hence make abstracts more accessible.

For example, the author index to Volume 69 (1968) of *Chemical Abstracts (CA)* was compiled, organized, and composed for printing almost entirely by computer. Most of the manual sorting, recopying, and checking of data ordinarily required in preparing an index was eliminated. It is significant to note that the same data need not be keyboarded and verified again for any other applications such as the five-year *Collective Index* to *CA*. As an example of the time savings that computers can achieve, *CA* officials hope to publish the 1967–1971 *Collective Index* within 30 months as compared with the nearly 48 months required to produce the previous *Collective Index* by manual methods.

In addition, some investigators believe that automatic (computer-based) document analysis and searching is not only as efficient as human indexing but may supplant it in future years. The questions of economics and of the availability of documents in machine-readable form will affect this possibility. A full discussion of this is beyond the scope of this book.

14.3. AUTOMATING THE ABSTRACTING PROCESS

14.3.1. Writing and Editing

Some abstractors now have available to them specific methods and devices that can (*a*) help the abstractor in the process of creating, revising, and editing the abstract and (*b*) simultaneously capture information in machine-readable form on punched paper or magnetic tape. These could be devices such as the "Friden 2340" or the IBM "Magnetic Tape Selectric® Typewriter (MTST)" units, which can produce machine-readable output. Since these are not directly connected to the computer they are called "off-line" devices.

Alternatively, the abstractor could use "on-line" devices (directly connected to the computer). These could be special typewriter terminals connected by a telephone wire from remote points to a computer that could service several such terminals at the same time and also do other work.

With such devices (whether on-line or off-line), once corrections are made, any further retyping or proofreading is eliminated because the original has been "captured" in magnetic or mechanical form. This reduces costs and lessens chances for error.

Also, if abstracts are available in machine-readable form as input to a computer, index terms could be extracted automatically. In advanced systems, such as that used by the IBM Technical Information Retrieval Center, the entire text of abstracts can be searched either retrospectively (for older material) or for current alerting, both without the necessity for human indexing.

14.3.2.   Automatic Abstracting

Although it appears unlikely that completely automatic abstracting based on the use of computers will replace human abstractors in the foreseeable future, this is a field in which useful work has been done in the past and is still being done. Most abstractors will want to follow this field closely.

Typically, at least in the past, such work has involved the extraction of "key" sentences based on the frequency of words occurring in the text of the entire document. More recent work, most notably that of James E. Rush and colleagues at the Ohio State University, has emphasized as goals the identification and elimination of sentences of low information value and the identification and inclusion of sentences of high information value. This preserves for the abstract a relatively small group of sentences that contain the essence of the information in the document. Examples of exclusions might include preliminary remarks, equations, footnotes and references, quotations, tables, charts, figures, graphs, speculative statements, opinions, and comparative clauses. Examples of sentences to be included would be those relating to objectives of the work, results, and conclusions.

Further progress in the area of automatic abstracting is directly related to the availability of original documents in machine-readable form. Such availability, in combination with novel approaches such as those of Rush, could have a dramatic impact on the success of further research in this area.

14.4.   INTEREST PROFILES AND SELECTIVE DISSEMINATION

An "interest profile" is a carefully organized reflection of the current interests of a user. Its purpose is to ensure that information center personnel send to the user those abstracts and other information most closely related to user interests. Dispensing information in this way is known as Selective Dissemination of Information (SDI) and is now usually done with a computer-based system.

Interest profiles can be expressed by clients as a list of key words either (a) based on a paragraph such as:

"I am interested in the British-French project on the supersonic transport (SST) aircraft. This craft, known as Concorde, is a joint venture of British Aircraft Corp. (BAC) and Sud-Aviation."

or (b) developed directly without the intermediate step (a). For example, the following list of keywords could be derived from the above paragraph:

1. SST.
2. Supersonic Transport Aircraft.
3. Concorde.
4. Sud-Aviation.
5. British Aircraft Corporation.
6. BAC.

The construction of interest profiles may seem easy—our examples were not complex—but in actual practice a great deal of care and precision in choosing the right words is required. In addition, "logic" can be introduced by specifying combinations of keywords or by specifying keywords which are not of interest. For example,

"I am interested in A in combination with B or C except that I'm not interested in any such combination which also includes D or E."

A good profile is so constructed that a minimum number of documents of possible interest will "escape." At the same time, some trimming is usually necessary so that clients will not be flooded with useless documents. To achieve these dual goals, the information specialist must review the words submitted for entry into the profile with great care. Once entered, profiles should be kept up to date as based on:

1. Development of new user interests.
2. Obsolescence of interests.
3. Evaluation by the user of the notices he gets (too many, too few, pertinency).

Lancaster (15) and others have discussed the question of walking the "tight rope" between providing too much and too little.

The computer searchable input into a computer-based SDI system is twofold:

1. Interest profiles of users.
2. Entries for abstracted documents.
    These can consist, for example, of (a) the complete entry including

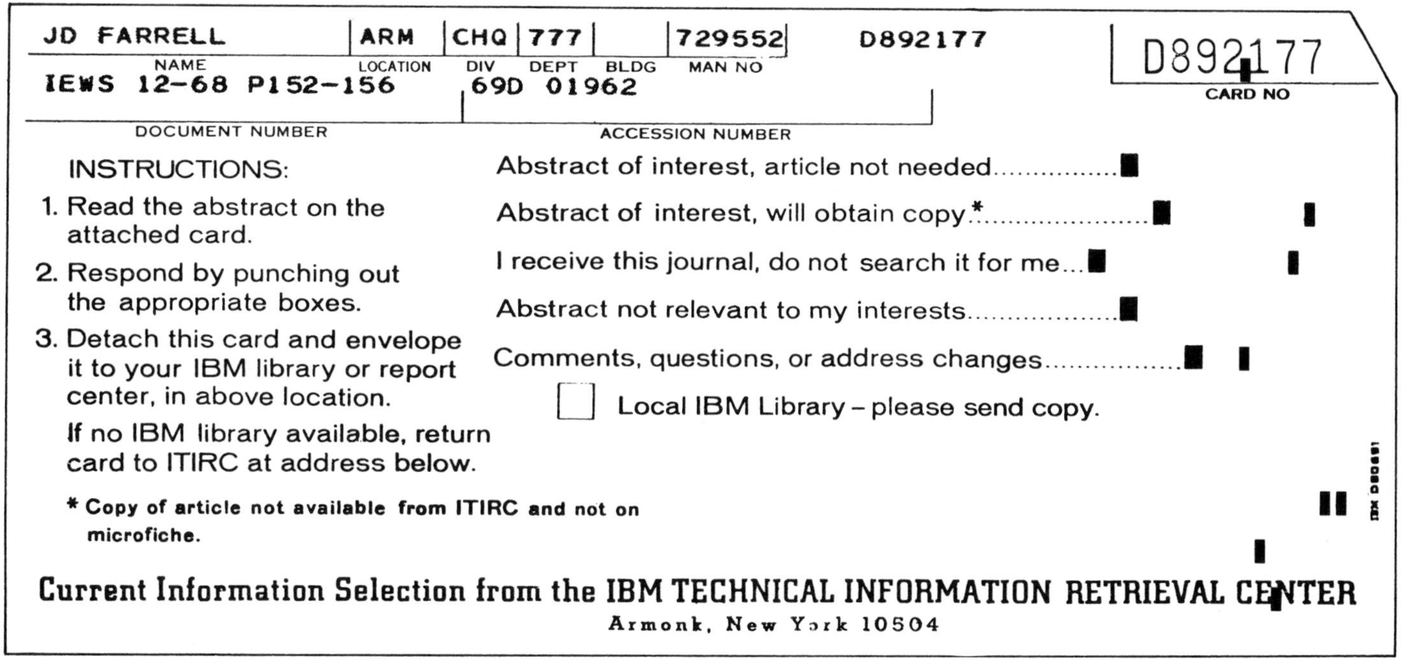

Fig. 14.1. Selective dissemination of information abstract notices. *Source.* J. J. Magnino, Jr, International Business Machines Corporation, Armonk, N.Y. Reprinted with permission.

The card contains:

JD FARRELL

| NAME | | ARM | CHQ | 777 | | 729552 | D892177 |

LOCATION: 12-68 P152-156

DIV: 69D DEPT: 01962 BLDG MAN NO

IEWS 12-68 P152-156

DOCUMENT NUMBER — ACCESSION NUMBER

D892177

CARD NO

INSTRUCTIONS:

1. Read the abstract on the attached card.

2. Respond by punching out the appropriate boxes.

3. Detach this card and envelope it to your IBM library or report center, in above location.

If no IBM library available, return card to ITIRC at address below.

* Copy of article not available from ITIRC and not on microfiche.

Abstract of interest, article not needed ....

Abstract of interest, will obtain copy* ....

I receive this journal, do not search it for me ....

Abstract not relevant to my interests ....

Comments, questions, or address changes ....

☐ Local IBM Library – please send copy.

Current Information Selection from the IBM TECHNICAL INFORMATION RETRIEVAL CENTER

Armonk, New York 10504

188

69D 01962   JOURNALS   03/17/69

IEWS 12-68 P152-156.   ADMINISTRATIVE TERMINAL SYSTEM FOR IBM SYSTEM/
360.   DECEMBER 1968.
IEEE TRANSACTIONS ON ENGINEERING WRITING AND SPEECH
FREEMAN, DE
IBM, WASHINGTON, D.C.
THE ADMINISTRATIVE TERMINAL SYSTEM FOR IBM SYSTEM/ 360 (ATS/
360) IS A NEW TIME SHARED DOCUMENTATION SYSTEM. THIS ON LINE SYSTEM
HELPS SOLVE PROBLEMS IN THE PREPARATION OF ENGINEERING DOCUMENTS BY
THE USE OF SPECIALLY DESIGNED TERMINAL COMMANDS FOR TEXT MATERIAL.
AND BY A PROGRAMMING SYSTEM TO PROVIDE MANY DOCUMENT PROCESSING
FUNCTIONS. THESE INCLUDE AUTOMATIC PAGE FORMATTING WITH HEADINGS.
FOOTINGS, AND PAGE NUMBERING, AS WELL AS TEXT MODIFYING ACTIONS. A
SIGNIFICANT FEATURE IS THE EASE WITH WHICH DOCUMENTS CAN BE CONVERTED
TO MACHINE READABLE FORM AT A TYPEWRITER TERMINAL FOR FURTHER
COMPUTER PROCESSING. PROGRAMMING TECHNIQUES USED ALLOW UNRELATED
BACKGROUND PROGRAMS TO RUN CONCURRENTLY WITH ATS/ 360. OBVIATING THE
NEED FOR A DEDICATED COMPUTING SYSTEM.

Fig. 14.1. (Continued)

69D 02359 JOURNALS 03/17/69

HUFA 08-68 P431-437. HUMAN FACTORS EVALUATION OF A COMPUTER BASED
INFORMATION STORAGE AND RETRIEVAL SYSTEM. AUGUST 1968.
HUMAN FACTORS
BARRETT, GV THORNTON, CL CABE, PA
GOODYEAR AEROSPACE CORP., AKRON, OHIO
THE HUMAN FACTORS ASPECTS OF A COMPUTER BASED INFORMATION
STORAGE AND RETRIEVAL SYSTEM WERE EVALUATED IN THREE GOVERNMENT
INTELLIGENCE SERVICES USING QUESTIONNAIRE AND INTERVIEW TECHNIQUES.
IT WAS FOUND THAT MANY FACTORS ENTERED INTO SATISFACTION WITH THE
INFORMATION SYSTEM, INCLUDING FAMILIARITY, EASE OF USE, AND
IMPORTANCE. OTHER RELEVANT FACTORS INCLUDED TRAINING IN THE USE OF
THE SYSTEM, AMOUNT AND TYPE OF INFORMATION TO MEET NEEDS IN THE
SYSTEM, AND TOLERANCES OF INDIVIDUALS FOR IRRELEVANT MATERIAL IN THE
OUTPUT OF SEARCHES. THE INTERACTION OF THESE FACTORS IS DISCUSSED IN
RELATION TO SATISFACTION WITH THE SYSTEM.

IBM TECHNICAL INFORMATION RETRIEVAL CENTER

IBM D60691

Fig. 14.1. (Continued)

DTMN 01-69 P87.   SOUTHERN PACIFIC SHOWS $22-MILLION TOPS INFORMATION
NETWORK.   JANUARY 1969.
     DATAMATION
     SOUTHERN PACIFIC BROUGHT OUT NOT ONLY ITS OWN TOP MANAGEMENT
BUT ALSO SUCH IBM NOTABLES AS EDP HEAD FG RODGERS AT A PRESS
CONFERENCE TO SHOW OFF TOPS (TOTAL OPERATIONS PROCESSING SYSTEM) AT
SP SAN FRANCISCO HEADQUARTERS.
     NO IN PARTIAL OPERATION AFTER 8-YEARS (OR ABOUT 600 MAN YEARS)
OF DEVELOPMENT, TOPS WILL USE TWO 360/ 65, THREE 360/ 40, FOUR 360/
20, SEVEN 2314 DISK UNITS, 200 1050 DATA COMMUNICATION SYSTEMS,
THIRTY TWO 2260 CRT DISPLAY TERMINALS, 200 TELEPRINTERS AND ROOMSFUL
OF SUPPORTING EQUIPMENT TO KEEP TRACK OF 2,300 LOCOMOTIVES AND 93,000
FREIGHT CARS SOMEWHERE OR OTHER ON THE RAILROAD'S 14,000 MILES OF
TRACK.   THE $14-MILLION WORTH OF IBM EQUIPMENT IS TIED TOGETHER BY A
PRIVATE MICROWAVE SYSTEM, BRINGING THE TOTAL COST TO ABOUT
$22-MILLION.

IBM TECHNICAL INFORMATION RETRIEVAL CENTER

IBM 080691

Fig. 14.1.   (Continued)

69D 02362

JOURNALS   03/17/69

CPAU 12-68 P25-29. AUTOMATED RETRIEVAL OF LEGAL INFORMATION— STATE
OF THE ART. DECEMBER 1968.
    COMPUTERS AND AUTOMATION
    FURTH, SE
    IBM, WHITE PLAINS, N.Y.
    AT THE PRESENT ADVANCED STAGE OF DEVELOPMENT IN THE COMPUTER
AND PROGRAMMING ARTS, AUTOMATED LEGAL OR STATUTORY SEARCH SYSTEMS ARE
CONVINCINGLY PROVING THEIR WORTH IN RELIABLE, TIME AND MONEY SAVING
DAILY USE.
    COMPUTER METHODS ARE DRAMATICALLY MEETING THE RAPIDLY
ACCELERATING NEEDS FOR QUICK, ACCURATE, THOROUGH METHODS OF FINDING
STATUTORY AND OTHER TEXTUAL DATA. THEY DO NOT RELY ON TRADITIONAL,
TIME CONSUMING, ERROR PRONE MANUAL RESEARCH WORK INVOLVING VOLUMINOUS
PUBLISHED WORKS, INDEXES AND CITATIONS.
    IN A FEW YEARS LEGISLATORS, GOVERNMENT AGENCIES AND LAWYERS
THROUGHOUT THE COUNTRY ARE LIKELY TO ACCEPT COMPUTERIZED SEARCH
SERVICES AS A STANDARD PROCEDURE.

Fig. 14.1. *(Continued)*

title, authors, reference, abstract text, and index entries or (*b*) title, authors, reference, and index entries, but not the abstract text.

When a word (or words) in a user profile is suitably matched with a word (or words) in an incoming (new) document entry, a so-called "hit" results. The computer then automatically prints an SDI notice intended for rapid dissemination to the user.

An outstanding example of an SDI system is the work of Magnino and co-workers at IBM (18). This involves "full text" searching (the complete abstract is searched by the computer) and resulting notices of hits also contain full abstracts as shown in Fig. 14.1. Another significant example of SDI is the "ASCA" service offered by the Institute for Scientific Information (Philadelphia).

14.5.   USE OF MAGNETIC TAPE EDITIONS OF ABSTRACTING SERVICES

Besides offering the advantage of rapid, automated searching for and locating of pertinent abstracts, the tape versions of abstract services often precede the printed versions by several weeks because of the time required to print, bind, and distribute the printed version.

A magnetic tape is a plastic tape coated with a magnetic material (such as ferrous oxide) on which information may be recorded by keyboard- or computer-driven devices. The information on the tape can be "read" (processed) by the computer. In abstracting operations a variety of tapes are now being produced as by-products of processing of data for the printed version.

Magnetic tape forms of (*a*) indexes to abstracts, (*b*) indexes and citations to abstracts, and (*c*) in some cases, both indexes and complete abstracts have become increasingly prevalent. Examples of these tapes, known as *data bases*, which have been made available by abstracting or abstracting-related services, include the following:

Basic Journal Abstracts (Chemical Abstracts Service)
BA Previews—includes *BioResearch Index* (BIOSIS)
CA-Condensates (Chemical Abstracts Service)
Chemical-Biological Activities (Chemical Abstracts Service)
Compendex (Engineering Index)
Excerpta Medica Information Systems, Inc. (3i Company/Information Interscience, Inc.)
Geo. Ref (American Geological Institute)

Index Chemicus Registry System (Institute for Scientific Information)
Inspec (Institution of Electrical Engineers)
Nuclear Science Abstracts (Atomic Energy Commission)
Polymer Science and Technology (Chemical Abstracts Service)
Searchable Physics Information Notices (American Institute of Physics)

The preferred method of working with the tapes is to use them in connection with current alerting (SDI) activities. The tapes are also used to a lesser extent for retrospective searching. The tapes can be searched in several ways:

1.   In some cases the abstracting services search their own tapes for users on a fee basis (often a basic annual contract plus a special fee for each search).

2.   Industrial firms and other organizations can buy or lease the tapes to do their own searching on site, using their own computers. The advantages of this are that interests can be kept proprietary and that the tapes can be manipulated to meet special needs.

3.   Searches of one or more of the tapes can be obtained through scientific information dissemination centers as described below.

Since the tapes are relatively expensive and require a high level of computer equipment and computer personnel skills, only the larger users can both afford to buy the tapes, and have the technical capability to work with the tapes.

The directors of the abstracting services have recognized this and also the fact that even large abstracting services may not have the resources to do very much mass processing of the tapes in response to user requests. As a consequence, most abstracting services that offer tapes have chosen to also make them available to the scientific community through scientific information dissemination centers—usually computer centers or information bureaus at universities or institutes such as the following:

Greater Louisville Technical Referral Center
Illinois Institute of Technology Research Institute (Chicago)
3i Company/Information Interscience, Inc. (Philadelphia)
University of Georgia
University of Iowa
University of Pittsburgh

Services available from these centers usually include: (*a*) current awareness and retrospective searching of the tapes from several suppliers for a fee based on the number or extent of interest profiles; (*b*) workshops on the use of the tapes; and (*c*) assistance in developing and maintaining profiles. Some of the centers (and other interested parties) have formed a group known as the Association of Scientific Information Dissemination Centers to share information and experiences and to promote applicable technology.

14.6. COMPUTER-BASED TYPESETTING

The abstractor who plans to work with one of the larger abstracting services should be aware that some of these services either already are using or plan to use computers for typesetting of some of their output.

One example of the use of computer-based typesetting is the three-part publication *Science Abstracts*, which is part of what is called INSPEC (Information Service in Physics, Electrotechnology, and Computers and Control). The three specific publications involved in *Science Abstracts* are *Physics Abstracts, Electrical and Electronics Abstracts*, and *Computer and Control Abstracts*.

Other examples include the American Geological Institute in its abstracting activities and the publications of the Excerpta Medica Foundation. Also, in its drive towards almost total computerization, the Chemical Abstracts Service is moving rapidly in this direction.

Probably the principal advantage claimed for computerized typesetting is that machine-readable output (such as magnetic tape or punched paper tape) can be obtained in connection with the printing process. This output can then be searched by computer as needed both to retrieve information and to generate a variety of special publications. Another advantage claimed is the large variety of characters—perhaps in the order of 1500—that may be available in a variety of sizes and typographic styles when computer-based systems are used.

A hypothetical sequence of steps in computerized typesetting might be approximately as follows:

1.   Keyboarding (typing) of the abstract using a device such as the "Friden Flexowriter" or IBM "MTST". This produces a punched paper tape or a magnetic tape (and typed copy which can be used to assist in proofreading if desired).

2.   Conversion of the resulting tape into computer-readable form, input into a computer such as the IBM "360/30" and production of printout.

3.   Proofreading of printout and rekeyboarding on the "Flexowriter" or "MTST" of parts that need correction.

4.   Input of corrected tape from step (3) into computer. Computer

processing: correction, hyphenation, justification, and formatting (including changes in type size and style).

5.   Use of revised tape from step (4) to operate an automatic composition device which produces camera-ready copy. This copy is readied for printing in a series of conventional steps.

6.   Magnetic tape generated by the computer as part of the process can be used for information retrieval or to produce specialized collections of abstracts and other products. *Thus, a single input yields multiple output.*

Walter (19) has reviewed the state of the art in computerized typesetting. Among the larger companies working on advanced concepts of computer-based typesetting are RCA, Mergenthaler Linotype, and IBM. Some of this work involves the use of television-like cathode ray tubes to help achieve high speeds.

The abstractor should make careful note of some of the potential disadvantages of computer typesetting. One is difficulty in handling mathematical expressions. Another is the necessity for fully operational and satisfactory computer programs and for sophisticated equipment, both of which can be expensive. Conventional typesetting may still be the route of choice for many organizations for years to come.

14.7.   SYSTEMS AND THE ABSTRACTOR

It is important for the abstractor to know how his work and its processing by computer fits into a working system.

14.7.1.   Larger Systems

The advantages to both abstractor and user of full scale computerization are most evident in large systems. This has best been described by Tate of Chemical Abstracts Service (20). According to Tate, in the computer-based system being planned by CAS, "a single keyboarding will put on tape all data selected in a single intellectual analysis of the source documents, an analysis combining both abstracting and indexing. We can then program to give us whatever product we want." Plans also include a composition system operated through the computer. The whole procedure, among other things, limits redundant effort and minimizes chance for error. In addition to the advantages in speed and economy made possible by computer, Tate points to the greater depth of indexing that is made possible when all words in the index entries and of the abstracts proper are in machine-readable form.

Figures 14.2 through 14.5 show the systems of CAS, the Defense Documentation Center, the National Aeronautics and Space Administration, and

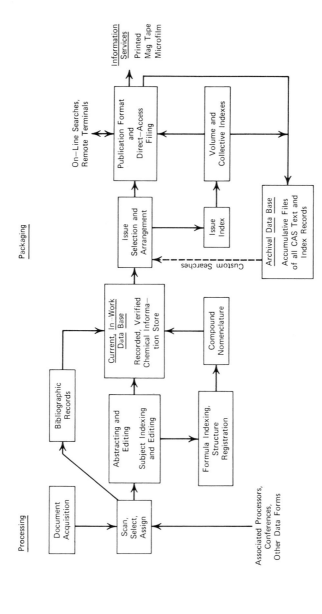

Fig. 14.2. Chemical Abstracts Service Data Base work flow. *Source.* Chemical Abstracts Service, American Chemical Society. Reprinted with permission.

197

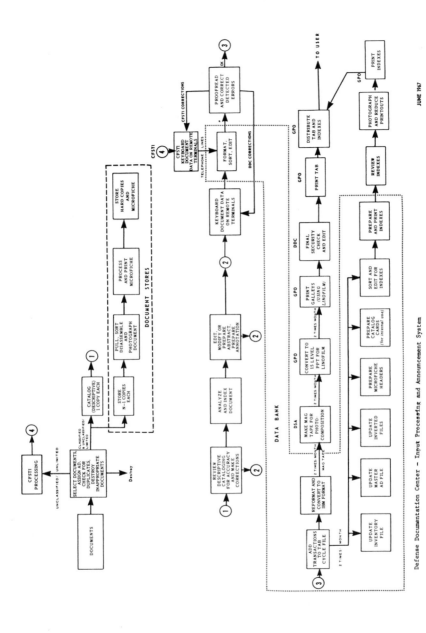

Fig. 14.3. Defense Documentation Center: system work flow. *Source. Selected Mechanized Scientific and Technical Information Systems*, Committee on Scientific and Technical Information (COSATI), Washington, D. C., April 1968.

Defense Documentation Center – Input Processing and Announcement System

198

Fig. 14.3. *(Continued)*

Defense Documentation Center – Bibliography Request Processing

JUNE 1967

199

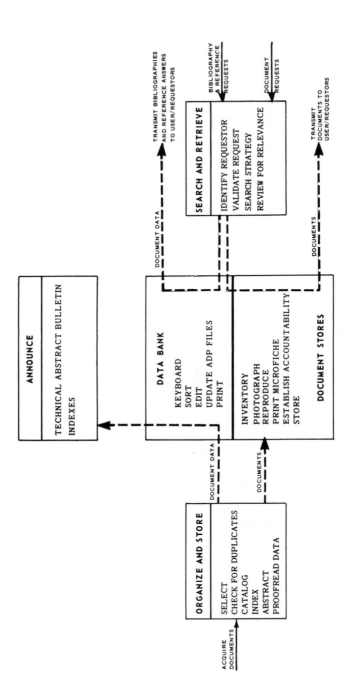

Defense Documentation Center – Overall System

Fig. 14.3. *(Continued)*

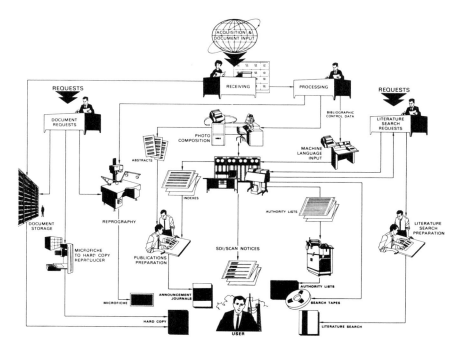

Fig. 14.4. National Aeronautics and Space Administration: information system work flow. *Source. Selected Mechanized Scientific and Technical Information Systems,* Committee on Scientific and Technical Information (COSATI), Washington, D. C., April 1968. Reprinted with permission.

the United States Geological Survey, and illustrate the work-flow in a computer-based system.

14.7.2. In-House Systems

In-house systems can also benefit by prudent computerization and can achieve good results. Thus, assuming that there are suitable programs and equipment, machine-readable input can be used to produce, at high speeds, a variety of output, as, for example, printing abstracts:

1.   In selected fields as part of the literature searching effort of an information center.
2.   On cards for use in information center files or library catalogs.
3.   On cards distributed to individual scientists as part of a current alerting program.
4.   As part of an accessions list or bulletin of current literature.

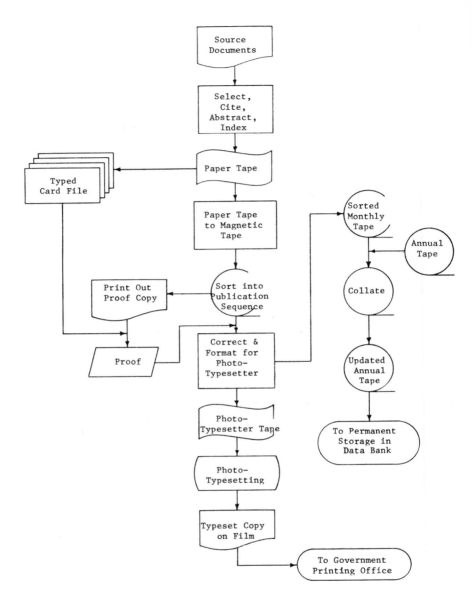

Fig. 14.5. U. S. Geological Survey: information system work flow. *Source. Selected Mechanized Scientific and Technical Information Systems,* Committee on Scientific and Technical Information (COSATI), Washington, D. C., April 1968. Reprinted with permission.

202

See Figs. 14.6 and 14.7 for examples of work flow in in-house systems. Another example is Fig. 14.8, which also shows the role of SDI within the total system.

14.8. INTERNATIONAL AND NATIONAL COOPERATION IN ABSTRACTING AND COMPUTER SYSTEMS

Use of the computer as a part of the abstracting process unlocks a vast potential for cooperation especially at the international level. For example, abstracts and index entries for documents in the field of chemistry prepared in West Germany and the United Kingdom could be (*a*) produced in

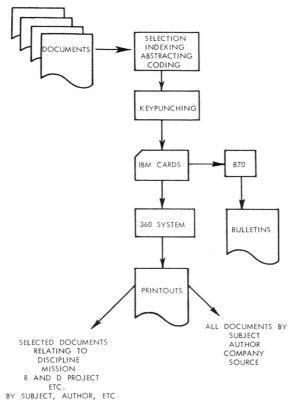

Fig. 14.6. Work flow in in-house systems: Hercules Technical Information Division. *Source.* H. Skolnik and R. E. Curtiss, "A Mechanized Information System For Many Outputs From One Input," *J. Chem. Doc.,* **8,** No. 1, February 1968, p. 45. Copyright American Chemical Society. Reprinted with permission of copyright owner.

machine acceptable form in these countries and then (*b*) centrally processed by the Chemical Abstracts Service (CAS) in the United States to produce output from which the German, British, and American chem-

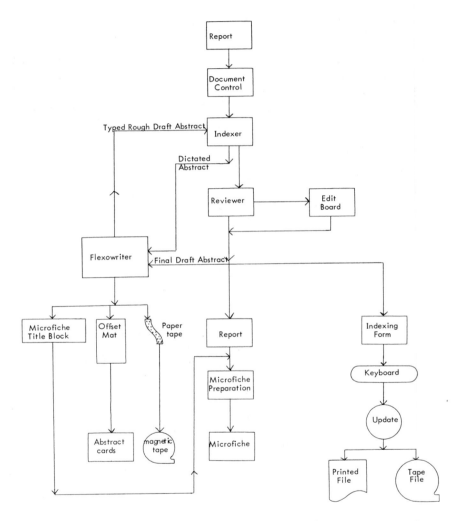

Fig. 14.7. Work flow in in-house systems: DuPont Central Report Index. *Source.* B. A. Montague and R. F. Schirmer, "DuPont Central Report Index," *J. Chem. Doc.,* **8,** No. 1, February 1968, pp. 37, 39. Copyright American Chemical Society. Reprinted with permission of copyright owner.

ical communites could derive specialized publications and services for users in these countries.

To achieve goals such as these, in 1969 CAS completed agreements

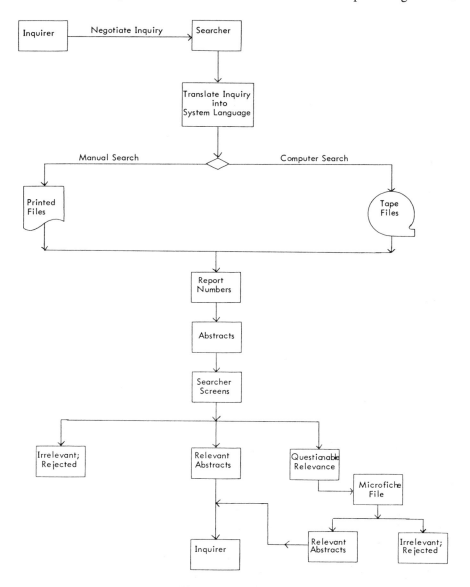

Fig. 14.7. *(Continued)*

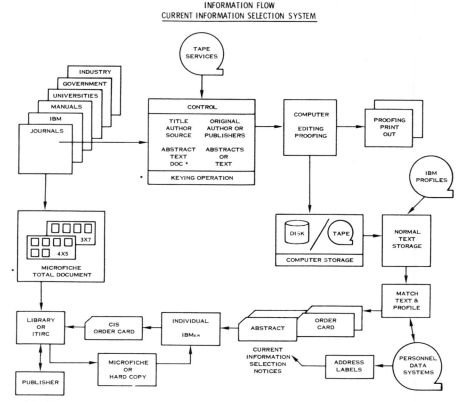

Fig. 14.8. IBM Current Information Selection System. *Source.* J. J. Magnino Jr., International Business Machines Corporation, Armonk, N. Y. Reprinted with permission.

with the United Kingdom Consortium on Chemical Information, acting through The Chemical Society (London), and with the Gesellschaft Deutscher Chemiker of West Germany. Additionally, in 1969, The Chemical Society (London) announced the establishment of a United Kingdom Chemical Information Service (with the principal center at the University of Nottingham) to provide information services using machine-readable input from CAS.

Other noteworthy efforts toward bringing about international cooperation include those of the International Council of Scientific Unions-Abstracting Board (ICSU-AB) and of the ICSU-UNESCO Central Committee. An example of the activities of the former is exploring the feasibility of standardizing input for abstracting systems; the latter group has been investigating the feasibility of a worldwide science information system.

Coordinated efforts toward international cooperation, if successful, could result in substantial improvements in the speed and accuracy of abstracting in chemistry, and other fields. There are many who believe that such cooperation is the only way for large abstracting services both to keep up to date and to maintain fiscal stability.

An example of *national* cooperation is found in the two complementary abstract journals which relate to the United States National Aeronautics and Space Administration (NASA) information system. These journals are:

> *Scientific & Technical Aerospace Reports,* published by NASA—covers report literature
>
> *International Aerospace Abstracts,* published by The American Institute of Aeronautics and Astronautics (New York)—covers trade journals, books, and papers presented at meetings.

Since the two publications are organized and indexed in basically identical ways, both can constitute sources of input into the NASA computerized information bank.

Computer-based systems can lead to additional beneficial relationships between primary journals and abstracting services. This concept is being explored actively by the American Chemical Society (21, 22) and by other groups. These studies could lead to reduced costs, increased timeliness, and improved utility. As one step in this direction, by the end of 1968, four American Chemical Society primary journals—*Journal of Chemical Documentation, Journal of Chemical and Engineering Data, Industrial and Engineering Chemistry–Process Design and Development,* and *Industrial and Engineering Chemistry–Product Research and Development* —were being composed through a computer-based system. This is part of planning for the possible development of a unified system for recording published information in machine-readable form as part of the publication process—a system which could be useful both to the primary journals and to the abstracting services.

Examples of what is being done, can be done, or is being considered include:

1. Use of the ASTM CODEN (unique five letter codes for journal titles published by the American Society for Testing and Materials) to validate journal citations and to reduce keyboarding. (The CODEN is maintained at Science Information Services, Franklin Institute Research Laboratories, Philadelphia.)

2.   Indexing chemical compounds directly from manuscripts accepted by primary journals.

3.   Preparation of volume indexes for primary publications by automatic extraction of index entries used to generate abstracting service entries.

4.   For selected primary journals, keying the abstracting service index entry directly to the original paper rather than to an abstract.

5.   Direct extraction of chemical structures: once keyboarded for the primary publication it would not need to be keyboarded again for the abstracting service.

At both the national and international levels, abstracting and indexing of abstracts could be expedited in other ways. For example, the author abstract and other selected information could be extracted for direct processing by an abstracting service. If an author abstract is not available, the machine-readable output for the entire document could be used to automatically generate an abstract in machine-readable form. Also, perhaps indexable words and concepts could be automatically identified and extracted for use in indexes of abstracting services.

14.9.   DIRECT COMPUTER-BASED ACCESS

The abstractor should be aware that during the 1970s users are expected to begin locating and using abstracts through remote terminals linked *directly* to a computer-based system. The results could be received by the user in printed, video (TV-like), or even audio form. Among the organizations working in this directing are NASA, the *New York Times,* and the Institute of Electrical and Electronics Engineers. The advantages to the user will be that he will be able to "converse" (communicate) directly with the computer with great speed.

14.10.   MICROFILMS AND ABSTRACTS

Microfilm technology is another field that the abstractor should try to follow since it can affect the way in which his product is used.

The printed versions of many abstract services are now widely available in 16-mm roll film in cartridge (cassette) form. Access to these cartridges is very quick and accurate by means of automatic readers which thread, wind, and rewind automatically. The readers are often printers as well, so that at a push of a button a print of an abstract of interest can be produced. Scientists can use computer and microfilm technologies together in this way:

1.   Request a computer search for a list of abstract numbers on a specific subject.

2.   Give the list of abstract numbers to a clerk who then prints the abstracts using an automatic microfilm reader-printer.

3.   Study the prints at one's desk to select original documents for detailed examination.

If the abstractor is in a position to do so, he should try to see that his abstracts are produced and microfilmed so as to assure maximum legibility and ease of use.

chapter 15

Evaluating Abstractors and Abstracting

By what criteria can the abstractor evaluate himself and the quality of his work? The criteria, specifically mentioned or alluded to elsewhere in this book, are summarized here. Under ideal conditions the following factors help to produce a quality abstract:

1. *Promptness* • Every effort is made to get the document to the abstractor as soon as possible. Once the original is received, the abstractor prepares the abstract as quickly as possible, at a maximum within 48 hours of receipt—unless overriding factors dictate otherwise. After the abstract is written, it is delivered to the user quickly—two to four weeks in the case of an express service.

2. *Accuracy* • The abstract is accurate both in content (especially quantitative or numerical data) and in reference to original source document. Accuracy is further enhanced by careful proofreading and by a minimum of repetitive transcription and keyboarding. If funds permit, the author of the original document or an editor provides an additional review.

3. *Clarity* • The abstract is clear and unambiguous. Trade names, jargon, acronyms, and the like are either adequately explained or not used at all.

4. *Readability* • The abstract is written in fluent, smoothly flowing English, and there is adequate punctuation. The style is concise but not awkward or stilted. The final product is easy to read by the user, both from the standpoint of *linguistics* and from the standpoint of *visual* appearance and clarity.

5. *Completeness* • The abstract is complete. No significant information is omitted. However, the abstract is not overloaded to the point where it is verbose and defeats the basic time-saving purpose of the abstract.

6. *Signing the Abstract* • The abstractor makes sure that his name appears after each abstract he writes or edits. The good abstractor takes pride in his work and deserves both full credit and full responsibility.

7. *Speculation and Criticism* • Any speculations of the author of the original document not based on fact are omitted. Also, if the abstractor is a competent specialist and is in a position to make any critical remarks in his abstracts, these remarks are clearly attributed to the abstractor.

8. *Client Needs* • The needs of the client are always kept in the foreground. These needs are satisfied by such means as orienting to meet specific interests. The abstractor tries to provide the amount of detail needed—not too much, not too little.

9. *Economy* • The abstract is produced at a minimum expense consistent with quality and accuracy. Full use is made of any good author abstracts that accompany the original document and of any material that can be properly excerpted from the original document.

10. *Selectivity* • The good abstract is intelligently written. The good abstractor does not insult the intelligence of the reader. He has insight into what is important and what is not important. He emphasizes the novel and the important and omits the obvious.

11. *Guidelines* • Although the abstractor prudently observes guidelines, he also works and writes in the manner and style with which he feels most comfortable and productive.

12. *Feedback* • Criticism is sought from other abstractors and from clients to sharpen abstract quality and to ensure that user needs are fulfilled. The abstractor is in close harmony with real needs.

Appendix A

Abstracting Scientific and Technical Reports

AD - 667 000

ABSTRACTING

Scientific and Technical Reports

of Defense-Sponsored RDT&E

March 1968

Defense Documentation Center

Cameron Station

Alexandria, Virginia 22314

Source. Reprinted with permission from Defense Documentation Center, Cameron Station, Alexandria, Va.; can be obtained from National Technical Information Service, 5285 Port Royal Road, Springfield, Va., 22151; cost $6.00.

Purpose

This publication provides guidance for the preparation of abstracts of technical reports. It is intended to point out the requirements an abstract must meet and to standardize its content and make it more adaptable to the automatic methods of information processing.

Scope

This publication is for use by persons required by their organization or by contract to prepare reports on the progress and findings of Government-sponsored research.

BY ORDER OF THE DIRECTOR, DEFENSE SUPPLY AGENCY

OFFICIAL

ROBERT B. STEGMAIER, JR.
Administrator
Defense Documentation Center

iii

"Write not that your reader may understand you if he wants to, but that he must understand you whether he wants to or not."

Foerster and Steadman
Sentences and Thinking

iv

216

Introduction

The word abstract is derived from the Latin *abstractus* meaning to draw from or separate. It is defined by Webster as "that which comprises or concentrates in itself the essential qualities of a larger thing or of several things." Therefore the abstract of a technical report should collect the essential aspects of the long paper and present them to the reader for his use.

With the volume of technical reports that is available to the scientist to accelerate his efforts in research and development, it is particularly important that he determine quickly the "essential qualities" of these reports and make a proper decision as to whether he should take the time to pursue their content in detail. The main purpose of the abstract should be to show the reader if he needs the full report by presenting a clear, concise, factual summary which is both an elaboration of the title and a condensation of the report.

The present trend toward automatic processing has created a second purpose for the abstract which may soon become the primary purpose. More and more the developers of automatic information processing techniques are relying on the abstract for their input and retrieval data. The reports are subject-indexed by the terms in the abstract based on their frequency, usage, and/or relative positions in sentences. If the abstract expresses the content of the document badly or inadequately the whole document and its valuable findings are lost.

-1-

Another major consideration of the abstract is its role in the current standardization programs. With the interchange of information being planned for complex communication networks, all contributors to this interchange should abide by certain standard rules so that all readers may understand and rely on what is written.

Based on all these facts, the following do's and don'ts are presented, not to change the style of any writer, but to try to make him aware of the requirements of a present-day abstract in the computer age and to let him know what is expected of him so that his contribution can be of optimum value to scientific and technical research and development.

Elements of Abstracts

An abstract should appear in every technical report. It should contain no more than 200-250 words, but these words individually and in combinations should reflect clearly, concisely and completely the subject of the report because they will probably be used to represent the report in an announcement journal or in a bibliographic listing, to anyone who reads the abstract or who wants to pigeonhole the report for future reference.

Since this abstract is to be a reflection of the substance of the report, it should be accurate and complete in itself. As much care should be given to writing the abstract as to writing the report because readers will decide from it whether they want to read the whole paper and because it may be used in lieu of the parent document for other purposes. It should be written in straightforward English making every word count. The sentences should be concise; subordinate clauses should be used only if they do not impair the ease of reading. To shorten, phrases should be used instead of clauses, and words instead of phrases, always keeping in mind that the final product must convey an accurate meaning.

Basically there are two kinds of abstracts, the descriptive and the informative. The descriptive abstract, as the name implies, is usually a statement of the general nature and scope of the report and should be used only when the complete document is such that it does not lend itself to an informative summarization. This type of abstract is usually supplied for bibliographies or collections of papers such as a symposium record.

-3-

More valuable is the informative abstract which succinctly informs the reader of the salient features of the subject studied. If properly written it reflects the organization of the report giving the objectives of the investigation, the methods employed in the research, and the conclusions reached and/or results obtained with special emphasis on any new data or dis-. coveries which would affect or be useful to other similar research or applications. These three elements can be expressed as follows:

- The objectives and/or purpose of the investigation are important because they set down the area under observation and immediately tell the reader whether the research concerns him. One brief sentence should be sufficient.

- The methods employed and the kind of treatment given -- mentioning materials, conditions, restrictions and limits -- should be pointed out specifically and selectively without going into detail.

- Results obtained and/or conclusions reached represent the essential part of the effort described in the document; therefore, most of the abstract should be devoted to this element. If the results include numerous specific data, a general statement should be made; or, if the scattered nature of data allows no such statement, this fact should be mentioned. If possible, all new findings should be clearly stated including any important numerical values. Conclusions which interpret the results should also appear.

-4-

Some documents, such as those on evaluation or qualification tests, can be abstracted by summarizing the results and recommendations for the disposition of the items. Abstracts of bibliographies should cite the scope of the work, time period covered, method of arrangement and the number of entries.

The terminology, particularly the technical terminology, in the abstract should be the same as that in the report. New concepts and procedures of automatic processing are utilizing the abstract for indexing. The technical terms which appear in the abstract may become the subject terms used for storage and retrieval of information about the report. If they are inadequate or different (e.g., optical masers for lasers) or more generic (e.g., steel for martensite) than those used in the document, they may not be effective if used to retrieve the document by subject content.

Some other factors should also be considered in preparing a good abstract. Unconventional or rarely used symbols and characters should not be used because they cannot be introduced into certain automatic data processing systems (see sample, Appendix I). In fact, considering the coming use of optical scanners for input processing, complete verbalization is desirable. Illustrations, preliminaries, descriptive details, examples, numbered equations and footnotes are to be avoided. Descriptive cataloging information should be omitted as, for example, corporate authors, contract numbers, report numbers, and reporting dates. Precise measurements should be expressed by numerals except at the beginning of a sentence; abbreviations, on the

-5-

other hand, should be limited to those in frequent use among professionals engaged in the research under discussion.

After the abstract is written to incorporate the elements enumerated above, it should be reviewed carefully and shortened through economy of phrasing and elimination of idle words. The final product should be approximately 200-250 words long (see examples, Appendix II).

Security Classification

It is highly desirable that the abstract be unclassified even if it summarizes a classified report. However, if it is necessary for complete and intelligible context to include specific data the unauthorized disclosure of which would be detrimental to the defense interests of the United States, such information must be classified. The classification guidance should reflect the latest applicable security regulations, as for example, DoD 5220.22M, "Industrial Security Manual for Safeguarding Classified Information," or DoD Instruction 5210.47, "Security Classification of Official Information." The abstract may be given a lower classification than the report, particularly if the results obtained can be reported in the abstract only in general terms. The security classification should never decide the nature of the abstract. It is better to have a classified abstract that is illuminating than to have an unclassified but ineffectual abstract.

Controlled Distribution Statements

To provide maximum information in the abstract, the author must sometimes include facts about proprietary matters and other interests of the issuing organization. To protect such information and prevent its unauthorized release through some announcement medium, a statement should be included in the report as to the restrictions to be imposed on the dissemination of the abstract per se. This is in addition to and distinct from any

-7-

limitation imposed on the report as a whole, and can be indicated on the DD Form 1473 (Document Control Data - R&D) which
must be incorporated into every technical report generated for
DoD (DoD Instruction 3200.8, "Standards for Documentation of
Technical Reports under the DoD Scientific and Technical Program"). It can be made as a parenthetical statement at the end
of the abstract until such time as other space is provided for
it. A distribution limitation, like a security classification,
should only be used when a specific reason for it exists.
Limited reports do not necessarily require limited abstracts.

Outline

In brief:

1. Always an informative abstract if possible
2. 200-250 words
3. Same technical terminology as in report
4. Contents
    a. Objectives or purpose of investigation
    b. Methods of investigation
    c. Results of investigation
    d. Validity of results
    e. Conclusions
    f. Applications
5. Numerals for numbers when possible
6. Phrases for clauses, words for phrases when possible
7. No unconventional or rare symbols or characters (see Appendix I, Verbalization Chart)
8. No uncommon abbreviations
9. No equations, footnotes, preliminaries
10. No descriptive cataloging data
11. Security Classification
12. Dissemination controls, if any
13. Review it.

APPENDIX I

VERBALIZING FOR MACHINABILITY*

The following symbols may be used in abstracts, annotations, and titles:

. , : ; ' / * $ % ( ) - + = < >

ANGSTROM UNITS ($\overset{\circ}{A}$)
Use A

CHEMICALS
H_2SO_4 *use* H2SO4

→ *use* yields

CUBIC
cm^3 *use* cu cm *or* cc

ft^3 *use* cu ft

m^3 *use* cu m

DEGREES
32° *use* 32 deg
32° F *use* 32 F
32° 16' 8" *use* 32 deg 16 min 8 sec

EXPONENTS
$x^{(n-1)}$ *use* x to the (n-1) power

ft sec^{-1} *use* ft/sec

When the exponent is less than 7 and has the base 10, write out the number; e.g.,

10^2 *use* 100

10^{-4} *use* 0.0001

2.75×10^{-3} *use* 0.00275

When the base is 10 and the exponent is 7 or more, write out; e.g.,

10^7 *use* 10 to the 7th power

10^{-9} *use* 10 to the minus 9th power

See also CUBIC, SQUARE, SUPERSCRIPTS

FRACTIONS
Use the slash (virgule) for the fraction bar; e.g.,

$x = \dfrac{a-b}{c}$ *use* x = (a-b)/c

$x = a - \dfrac{b}{c}$ *use* x = a-(b/c)

GREATER THAN OR EQUAL TO ($\geq$)
Use > or =

GREEK LETTERS
Use their names; e.g.,
α *use* alpha
β *use* beta
π *use* pi

ITALICS
Do not use; see also UNDERSCORING

LESS THAN OR EQUAL TO ($\leq$)
Use < or =

LOGARITHMS
$\log_{10}$ *use* log
$\log_e$ *use* ln

MATHEMATICAL SYMBOLS
See SPECIAL SYMBOLS

MICRO- and MICROMICRO-
μv *use* microvolts
$\mu\mu f$ *use* micromicrofarads or picofarads

MICRONS
μ *use* micron
$m\mu$ *use* millimicron
μm *use* micrometers

PLUS OR MINUS ($\pm$)
Use + or −

QUESTION MARK (?)
Do not use; rephrase sentence

QUOTATION MARKS (")
Use the apostrophe or single quote only; e.g., The term 'overkill'
Where quotation marks are conventionally used as a symbol, abbreviate; e.g.,
12" *use* 12 in
45" *use* 45 sec
See also DEGREES

*Standard Operating Procedures; DSA, Defense Documentation Center, Alexandria, Va.

APPENDIX I

VERBALIZING FOR MACHINABILITY*

SPECIAL SYMBOLS

$\left.\begin{array}{l}\cong\\\approx\end{array}\right\}$ *use* approximately

$\rightarrow$ $\left\{\begin{array}{l}use \text{ yields } (chemistry)\\use \text{ approaches limit of } (mathematics)\end{array}\right.$

\# *use* no.

& *use* and *in titles, abstracts, and annotations*

∞ *use* infinity

λ $\left\{\begin{array}{l}use \text{ wavelength } (electronics \ and \ physics)\\use \text{ lambda } (all \ other)\end{array}\right.$

Ω $\left\{\begin{array}{l}use \text{ ohms } (electricity \ and \ electronics)\\use \text{ omega } (all \ other)\end{array}\right.$

ϕ $\left\{\begin{array}{l}use \text{ phase } (electricity \ and \ electronics)\\use \text{ phi } (all \ other)\end{array}\right.$

Similarly, spell out or show by acceptable alphanumeric characters increment, varies as, therefore, differential of, variation of, integral, sum, benzene ring, thunderstorm, male, female, fixed star, *etc.*

SQUARE

cm^2 *use* sq cm

ft^2 *use* sq ft

m^2 *use* sq m

SQUARE ROOT

$\left.\begin{array}{l}\sqrt{a\text{-}b}\\(a\text{-}b)^{1/2}\end{array}\right\}$ *use* square root of (a-b)

SUBSCRIPTS

V_1 *use* V sub 1

B_5 *use* B *(omit the 5, which is the atomic number of boron)*
See also CHEMICALS

SUPERSCRIPTS

H^+ *use* H (+)

SO_4^{--} *use* SO4 (−)

V^{5+} *use* V(5+)

U^{234} *use* U234

B^{10} *use* B10

$O^{18}(p,n)N^{15}$ *use* O18(p,n)N15

d_{23}^{25} *use* density at 23 deg F referred to water at 25 deg F

n_D^{20} *use* index of refraction for 20 deg F and sodium light

See also CUBIC, EXPONENTS, SQUARE

UNDERSCORING

Do not use underscoring

Escherichia coli *use* Escherichia coli

to set off special terms use single quotes;
e.g.,

the term quasar *use* the term 'quasar'

---

*Standard Operating Procedures; DSA, Defense Documentation Center, Alexandria, Va.

APPENDIX I
(Page 2)

APPENDIX II

SAMPLE ABSTRACTS

INFORMATIVE ABSTRACTS:

AD-815 700

A scaled model test was conducted to determine the extent of asymmetrical nozzle flow separation during staging of the Minuteman missile. Tests were conducted at a constant interstage pressure of 20 psig, where possible, and at various stage separation distances using cold air. Configurations tested were as follows: obstruction simulating instrumentation package at the nozzle exit; simulated 1-2 interstage quadrants; and simulated 2-3 interstage quadrants. Test data indicated that there was little or no asymmetrical separation during staging.

AD-815 770

The objective of this test was to investigate the use of auxiliary parachutes to soften the impact shock of the conventional Parachute Low Altitude Delivery System. Photo theodolite cameras were used to measure the impact velocities of PLADS loads with and without auxiliary parachutes. Auxiliary G-13 parachutes were tested with 600 and 1200 lb loads. The impact velocity with the 1200 lb loads was reduced from 60 FPS to 40 FPS. The reduction achieved with 600 lb loads was negligible. It is recommended that loadmasters receiving C-123 training from the 4410th Combat Crew Training Wing be made familiar with this technique.

DESCRIPTIVE ABSTRACTS:

AD-644 689

This document is an annotated bibliography of the journal and technical report literature (to August 1966) on electromagnetic wave propagation in conducting media, with emphasis on the ocean. References on antenna theory and performance are also included. This work includes 164 references and an author index.

AD-643 937

The report reviews Soviet and Soviet-bloc laser research as reflected in the open scientific literature published in the USSR and the bloc countries. It is the second in the series and is based on 634 research and review papers which cover the period from September 1964 through February 1966. The first appeared as ATD Report P-65-23 (AD-615 177) and covered the period from January 1961 through August 1964. The review is divided into sections, each section covering the work of a particular organization (a research institute, a laboratory, or a university). These sections are, in most cases, subdivided into specialized subject areas. The review sections are arranged in the order in which they are discussed in the sections and are keyed to the appropriate review section by letter designation.

APPENDIX III

REFERENCE CONSULTED

1. Army Regulation 70-31, "Standards for Technical Writing," Headquarters, Dept. of the Army, Washington, D. C. 9 September 1966

2. "Style Manual," American Institute of Physics, New York. Second Edition, 1959

3. "ASTIA Guidelines for Cataloging and Abstracting," Robert L. Murphy, Armed Services Technical Information Agency (Now Defense Documentation Center), Alexandria, Va. June 1962

4. AD-157 134 "Report Writing Guide for Arnold Engineering Development Center (Revised Edition)," C. B. Kestler, Arnold Engineering Development Center, Tullahoma, Tenn. May 1958

5. "Writing at BRL," Ballistics Research Laboratories, Aberdeen Proving Ground, Md. July 1961

6. "Standard Format for Scientific and Technical Reports Prepared by Contractors or Grantees," COSATI, Federal Council for Science and Technology, Washington, D. C. June 1967.

7. PB6, "Style Manual for Research and Development Technical Publication," Frankford Arsenal, Philadelphia, Penna. February 1959

8. "Writing Scientific Papers and Reports," W. Paul Jones, Wm. C. Brown Co., Dubuque, Iowa. Third Edition, April 1957

9. "National Bureau of Standards Publications and Reports Manual," Thos. G. Hereford, U. S. Dept. of Commerce, National Bureau of Standards, Washington, D. C. June 1957

10. "Report Writer's Guide," Donald E. Thackrey, Willow Run Laboratories, University of Michigan, Ann Arbor, Mich. February 1960

11. AD-622 944, "Suggested Criteria for Titles, Abstracts and Index Terms in DoD Technical Reports," A. G. Hoshovsky, Office of Aerospace Research, U. S. Air Force.

APPENDIX III

DOCUMENT CONTROL DATA - R & D

(Security classification of title, body of abstract and indexing annotation must be entered when the overall report is classified)

1. ORIGINATING ACTIVITY (Corporate author)	2a. REPORT SECURITY CLASSIFICATION
Defense Documentation Center Cameron Station Alexandria, Virginia 22314	Unclassified
	2b. GROUP

3. REPORT TITLE

Abstracting Scientific and Technical Reports of Defense-sponsored RDT&E.

4. DESCRIPTIVE NOTES *(Type of report and inclusive dates)*

5. AUTHOR(S) *(First name, middle initial, last name)*

6. REPORT DATE	7a. TOTAL NO. OF PAGES	7b. NO. OF REFS
March 1968	18	

8a. CONTRACT OR GRANT NO.	9a. ORIGINATOR'S REPORT NUMBER(S)
b. PROJECT NO.	AD-667 000
c.	9b. OTHER REPORT NO(S) (Any other numbers that may be assigned this report)
d.	

10. DISTRIBUTION STATEMENT

This document has been approved for public release and sale; its distribution is unlimited.

11. SUPPLEMENTARY NOTES	12. SPONSORING MILITARY ACTIVITY

13. ABSTRACT

Abstracts of scientific and technical reports must be factual, accurate, clear, and concise so that they can effectively serve the research and development community in the present-day computer age. They must apprise the scientist of the content of the parent document; present to the documentalist appropriate technical terminology for his automatic processing of input and retrieval data; and provide a standardized item of information for easy interchange by the complex communication networks among the members of the R&D community. Abstracts may be informative or descriptive as needed. However, the preferred are the informative which give the objectives or purpose of the research, the methods of investigation, and the results and conclusions. Whenever possible, abstracts should be unclassified with no controls on their distribution.

DD FORM 1473
1 NOV 65

230

14. KEY WORDS	LINK A		LINK B		LINK C	
	ROLE	WT	ROLE	WT	ROLE	WT
*Abstracts - Processing Information retrieval Instruction manuals Subject indexing Reports Documentation						

Appendix B

Engineering Index

— Selected Pages of Abstracts

and punched on Teletype tape for transmittal to company and servicing railroad.   09633

Unit Trains

Unit trains go west; D.JACKSON,Jr; Coal Age v 74 n 9 Sept 1969 p 62-7; Entire output of Sunnyside, Utah, coking coal mine, owned by Kaiser Steel Co, of 5500 tpd is moved to Fontana, Calif, steel plant over distance of 806 mi (roundtrip distance 1612 mi) in 96-hr cycle in unit trains, consisting of 84 each 100 ton capacity cars; trains are loaded at mine from over 70,000 tons capacity stockpile at rate of 11,000 tph, that is highest in world, by gravity while being moved through 18 ft wide, 25 ft high, and 458 ft long tunnel, built under stockpile; two loading gates operators and one supervisor constitute loading crew.   09367

COAL WASHING. See Coal Preparation.

COBALT AND ALLOYS

See also Carbides; Catalysts—Cobalt; Films—Magnetic; Platinum Cobalt Alloys.

Corrosion

Corrosion of cobalt and cobalt alloys; F.R.MORRAL (Battelle Memorial Inst, Columbus, Ohio); Corrosion v 25 n 7 July 1969 p 307-22; Investigations are reviewed of composition and structures of corrosion films and scales found on cobalt and on simple complex cobalt-base industrial alloys; corrosion resistance of these films and scales depends on metal and alloy composition, environment, pressure, temperature, and time, additional factors, such as surface preparation, stress, thermal shock, geometry of part, velocity, density, and flow pattern of corroding environment also have effect; mechanisms of corrosion proposed are only hypotheses. 91 refs.   09679

Heat Treatment. See Cobalt Metallography.

COBALT CHROMIUM ALLOYS. See Gas Turbines—Materials.

COBALT METALLOGRAPHY

Contribution to knowledge of WC-precipitation hardenable Co-Cr-W-C alloys, (Beitrag zur Kenntnis WC-ausscheidungshaertbarer Co-Cr-W-C-Legierungen); O.KNOTEK (Soc des Soudures Castolin S.A., Lausanne, Switzerland), H.SEIFAHRT; Haerterei-Technische Mitteilungen v 24 n 1 Apr 1969 p 23-6; In metallographic study of cobalt base alloys with 2% C and 15 to 40% (W plus Cr), M_6C was found to form in addition to other complex carbides; formation of WC from M_6C (or eta-one phase) by heat treatment is described; M_6C is stabilized by increasing Cr and requires minimum temperature for decomposition; alloys with most of carbon precipitated as WC can be obtained. In German.   12859

COBALT ORE TREATMENT. See Metallurgy.

COKE

See also Adsorption; Chemical Plants—Power Supply; Electrodes —Carbon.

Reactivity of coke from unblended coal, green oil coke, or pitch; T.KASHINO, A.WASA, T.YOKOYAMA, M.MATSUI; Fuel Soc Japan—J v 48 n 502 Feb 1969 p 103-12; Test method for reactivity of fine coke is described in which apparatus is operated semiautomatically; coal coke decreases while rising carbonization temperature, reactivity of green oil coke, has tendency to increase; coal cokes with about 30% volatile matter have generally low reactivity; among coals with equal volatile matter, coke from coal with lower fluidity and dilation tends to have higher reactivity; among cokes produced from coals at same carbonization temperature, reactivity of coke becomes higher with increasing internal surface area of coke. 25 refs. In Japanese with English summary. 11779

Novel process makes coke from coal tar pitch; R.REMIREZ; Chem Eng v 76 n 4 Feb 24 1969 p 74-6; Plant for production of coke from coal tar pitch went onstream last fall in Tobata, Japan; plant is owned by Nittetsu Chemical Industries, subsidiary of Yawata Iron and Steel Co; plant basically converts pitch—low-value, large-volume byproduct of steelmaking—into high-grade product that competes effectively with petroleum coke; at Tobata, delayed-coking facility adjoins Yawata's steelworks, which pipe in coal tar pitch at about 302 F; this charge is first preheated by heat exchange with heavy oil; pitch is further heated by heat exchange with recycle oil from coke drums; pitch goes on to gas- or fuel oil-fired heater, where feed is partially cracked; process data and slowsheet. 11855

Formation of porous texture in coke during steam gasification, (Developpement de la texture poreuse des cokes de houille au cours de la gazeification par la vapeur d'eau); P.CHICHE (Centre d'Etudes et Recherches des Charbonnages de France, Verneuil-en-Halatte); J.COUE, S.DURIF, S.PREGERMAIN; Carbon v 7 n 2 Apr 1969 p 297-305; By measuring adsorption, densities, mercury penetration under pressure and small angle x-ray scattering, changes in porous texture of two cokes produced from high and low volatile coal during gasification with steam at 900 C were followed; study was also made of considerable differences in initial porosity of cokes, caused by preliminary treatment of raw material which was first finely ground and subsequently briquetted. 12096

Measurement of coke strength at temperatures up to 1500 C by drum test; K.BRADSHAW (Coalite and Chemical Products, Ltd, Chesterfield, England), H.C.WILKINSON; Inst Fuel—J v 42 n 338 Mar 1969 p 112-17; Review of measurements of strength of coke at elevated temperatures and description of tests with apparatus suitable for use at temperatures similar to those to which coke is subjected in industrial processes such as ironmaking; it is shown that there is no systematic relationship between strength of coke measured at high temperatures and that measured by conventional tests at normal temperatures, and that strength generally begins to deteriorate when testing temperature exceeds temperature of carbonization; different cokes display different behavior under experimental conditions. 24 refs. 12874

Briquetting

Studies on hot briquetting of char obtained by fluidized heat treatment of fine coal; H.NISHINO (Industrial Development Laboratory, Hokkaido, Japan), S.TAKEDA; Fuel Soc Japan—J v 48 n 504 Apr 1969 p 226-34; In study of manufacturing process of smokeless domestic fuel from Japanese fine coals of weakly caking property, by heat treatments at 400 to 475 C with fluidized bed, followed by direct briquetting of hot chars formed; when weakly caking coals of free swelling number ranging from 3 to 6 1/2 were used, they gave briquettes of high compression strength; properties of briquettes were good, when range of heat treatment temperature was between 325 and 450 C, and also when temperature of mold was maintained above 350 C. 22 refs. In Japanese with English summary. 11854

Metallurgical. See Blast Furnace Practice; Coal Preparation—Pelletizing.

Moisture. See Blast Furnace Practice—Control.

COKE MANUFACTURE

Studies into coking of brown coal low-temperature tar according to delayed coking principle—1, (Untersuchungen ueber die Verkokung von Braunkohlenschwelteeren nach dem delayed-coking-Prinzip); H.SCHMIERS (Inst fuer Organische Chemie der Bergakademie Freiberg, West Germany); Brennstoff-Chemie v 50 n 3 Mar 1969 p 78-82; Use of small semicommercial unit, pressure and temperature effect on coke and cracking products yield to study in pressure range from 2 to 6 atm and temperature range between 420 and 470 C; optimum conditions were established as 440 C and 4 atm (coke yield 19.5%); distribution of cracking product showed dependence not only on coking temperature but also on temperature gradient in coke tower. 13 refs. In German.
11776

Arsenic in coal charge, pitch, coke, and waste gases of coke oven batteries, (Arzen v koksu); V.MASEK (Klement Gottwald Steelwork, Ostrava-Kuncice, Czechoslovakia); Hutnicke Listy v 24 n 5 May 1969 p 323-5; Measurements show that some of originally very low arsenic content of coking coal is volatilized by high temperature heating; this is true to even larger extent for arsenic content of pitch; arsenic content of ash is therefore low, but flue dust contains up to 28 times the amount of arsenic in coal and up to 182 times that in pitch. In Czech.
12136

COKE OVENS

See also Industrial Wastes—Phenols.

Study into possibility of increasing coke oven efficiency, (Untersuchungen zur Leistungssteigerung des Koksofens); H.ECHTERHOFF (Bergbau-Forschung GmbH, Essen-Kray, West Germany), G.NASHAN; Brennstoff-Chemie v 49 n 12 Dec 1968 p 364-8; Extremely high coking temperatures were used in semicommercial installation to determine coke output under such conditions; temperatures applied were 1550, 1300, 1100 and 900 C, respectively; time saving up to 37% is possible (1550 to 900 C); coking process was anticipated for those temperatures in following relationships—volatile matter, coke strength and output vs coking time; coke properties did not deteriorate despite substantial decrease in coking time. In German.
11656

Choice of refractory materials for coke roasting; Yu.M.ABYZGIL'DIN, Z.I.SYUNYAEV; Chem & Technology of Fuels & Oils (English translation of Khimiya i Tekhnologiya Topliv i Masel) n 11-12 Nov-Dec 1968 p 807-10; Refractories with high SiO_2 content are not suitable for coke refining ovens because SiO_2 reduces to SiO which then escapes; refractories with high Al_2O_3 content should be used in construction of coke roasting ovens; refractories containing magnesium, calcium, and chromium oxides and which are likely to crumble during coke refining are not suitable to line walls of roasting kilns in coke desulfurization plants. 22 refs.
12292

Combustion

Estimation of volume of gas leaking from chambers to flues of coke-oven battery; A.R.JACK (British Coke Research Assn, Chesterfield, England); Inst Fuel—J v 42 n 341 June 1969 p 250-3; Volume was estimated by means of sulfur balance between fuel gas, waste gas and crude gas; rate of leakage from chambers of

COKE OVENS

battery as whole was about 4 cu m/100 cu m of fuel gas burnt which was equivalent to about 238 cu m/hr or 2.0% of total gas make. 10753

COKE PLANTS. See Industrial Wastes—Phenols.

COLLOID CHEMISTRY. See Polymers—Analysis.

COLOR. See Wool.

COLORIMETRY

Chromatic measurement; S.L.HERSH; Indus Photography v 18 n 1 Jan 1969 p 24-5, 52, 69-73; Tri-color system of eye is discussed, along with adaptation of eye to color; effect of local color adaptive effect; assessing color negatives; color balance and densitometry. 09580

COLUMBIUM AND ALLOYS

See also Molybdenum Columbium Alloys.

Creep

High temperature creep of certain columbium alloys, (Observations sur le comportement au fluage a haute temperature de certains alliages de niobium); J.MOULIN (Centre de Recherches du Groupe Pechiney, Voreppe, France), R.COUSSERANS, E.AL-HERITIERE, R.SYRE; Memoires Scientifiques de la Rev de Metallurgie v 66 n 3 Mar 1969 p 227-32; Study of effects of heat treatment and method of production on creep of Cb-10W-3Zr-0.3C alloy; results show that vacuum heat treatment at very high temperature has favorable effect because carbides dissolve and reprecipitate; more finely distributed carbides are identified; melting in electron beam furnace results in finer and more uniform structure than arc furnace melting does. In French. 11666

Electric Properties

Quench-aging of niobium-hydrogen alloys at -196°C; Y.SASAKI, M.AMANO; Japan Inst Metals—Trans v 10 n 1 Jan 1969 p 29-33; Decrease in electrical resistivity of columbium-hydrogen alloys is observed during isothermal aging at -196 C; it is shown that aging process is reaction with time exponent of 1/3 and activation energy of 0.10 plus or minus 0.02 ev; change in flow stress is also observed with aging at this temperature; these phenomena are not found in hydrogen-free columbium; it is suggested that changes in resistivity and flow stress should be attributed to hydride precipitation at -196 C. 15 refs. In English. 10838

COLUMBIUM CARBIDE. See Carbides.

COLUMNS

See also Beams and Girders; Elasticity; Structural Design.

Concrete

Analysis of reinforced concrete columns subjected to longitudinal load and biaxial bending; A.FARAH, M.W.HUGGINS; Am Concrete Inst—J v 66 n 7 July 1969 p 569-75; Integration method used in analysis leads to three simultaneous nonlinear equations which are solved by procedure based on Newton-Raphson method; example is solved and results are compared with those obtained from experiment; agreement is exceptionally good; integration method can be used for sections made of materials other than concrete provided that materials characteristics are defined. 12217

Tests of spirally reinforced concrete columns under short and long term centrical loads, (Versuche an wendelbewehrten Stahlbetonsaeulen unter kurz- und langzeitig wirkenden zentrischen Lasten); S.STOECKL, H.RUESCH; Munich.Technische Hochschule—

Materialpruefungsamt fuer das Bauwesen—Bericht n 67 1968, 44 p; Comparative test results of columns with spiral and vertical reinforcement, columns with spiral reinforcement only, and columns not reinforced; all columns had diameter of 15 cm and height of 60 cm; strength of concrete used was approximately 225 kp/sq cm, test age about 28 days; concrete covering of spiral was 0.5 cm, core diameter aimed at 13.5 cm; horizontal and vertical deformations were measured to after exceeding maximum load. 35 refs. In German with English summary. 12245

Stresses

One-way buckling of elasticity restrained columns; T.ALLAN (Univ of Strathclyde, Glasgow, Scotland); J Mech Eng Science v 11 n 3 June 1969 p 269-80; Effect of elastic lateral restraint is considered and analysis of long thin beam subject to point loading and of long thin column subject to axial loading is presented; this shows that behavior in such cases is similar and rather unique in that single discrete wave or buckle forms in member; axial load at which buckling occurs is shown to be independent of amplitude of buckle wave; experimental investigation is also reported which is in general agreement with analysis. 09231

Vibrations

Determination of fundamental natural frequencies of axially loaded columns and frames; J.NEOGY (Indian Inst of Tech, Kharagpur), M.K.S.MURTY; Instn Engrs (India)—J v 49 n 5 pt CI 3 Jan 1969 p 206-12; Characteristic equations of vibration problem were developed theoretically; exact solutions for frequencies of columns and frames are obtained by solving frequency equation with help of IBM 1620 digital computer using FORTRAN language; it is shown that there is practically linear relationship between axial load and corresponding column frequency and ratio of tangent modulus of elasticity to constant modulus of elasticity of material; use of natural frequencies of axially loaded members in determination of their buckling loads is indicated. 10827

Parametric response spectra for imperfect columns; M.L.MOODY (University of Colorado, Boulder); US Naval Research Laboratory—Shock & Vibration Bul 39 pt 3 Jan 1969 p 247-52; Response spectra for the maximum amplitude of mid-span deflection for a simply supported column subjected to 20 cycles of harmonically vary-axial force are presented. The nonlinear effects of inertia and large deflections are included in the investigation, and typical linear steady state solutions are shown for comparison. 12360

COMBINES. See Agricultural Machinery—Combines.

COMBUSTION

See also Carbon Monoxide—Combustion; Coal—Combustion; Coke Ovens—Combustion; Diesel Engines—Marine; Fuels—Combustion; Gas Engines—Combustion; Internal Combustion Engines—Combustion; Liquid Fuels.

Theory of stable homogeneous combustion of condensed substances; A.G.MERZHANOV (Inst of Chemical Physics, Moscow, Soviet Union); Combustion & Flame v 13 n 2 Apr 1969 p 143-56; Paper attempts to present systematic theory of stable homogeneous combustion of condensed substances (CS), generalizing some results obtained mainly in research work of present author and co-workers; elementary models of CS combustion are systematized and analyzed; staged combustion of CS and interaction of elementary mechanisms are discussed; classification of CS by their combustion mechanisms is proposed; certain experimental results are analyzed theoretically. 47 refs. 13348

COMMUNICATION. See Radio Telegraph; Telephone.

COMMUNICATION SATELLITES

Ultimate communications capacity of geostationary-satellite orbit; J.K.S.JOWETT (Space Systems Div of Post Office Headquarters, London, England), A.K.JEFFERIS; Instn Elec Engrs—Proc v 116 n 8 Aug 1969 p 1304-10; Geostationary-satellite orbit is useful for communication satellites; there is limit to number of communications channels which orbit can support in given bandwidth; capacity of $10°$ arc of orbit and 500 MHz bandwidth is determined as function of earth-station-antenna diameter and satellite-transmitter power; results for FM and PCM for telephony, and for FM only for TV; effect on results of variations in basic assumptions; these include interference noise permitted, interleaving of channel carrier frequencies, use of narrow-beam satellite antennas, and use of frequency bands higher than those at 4 GHz and 6 GHz. Paper 5912 E.
10498

The transmission installations of the Raisting ground radio station for communication via the Intelsat III satellites, (Die Sendeeinrichtungen der Erdefunkstelle Raisting fuer den Nachrichtenverkehr ueber die Intelsat-III-Satelliten); H.BAUM; Frequenz v 23 n 8 Aug 1969 p 226-32; Following a brief outline of communications effected so far with the aid of the synchronous Intelsat I and II satellites, the paper delineates the considerably expanded transmission possibilities afforded by the new Intelsat III satellites. Subsequently describes the transmitting facilities newly developed for the Raisting earth radio station of the German Federal Post, with particular emphasis on the related modem facilities for transmission of carrier multiplex, TV picture and TV sound signals. In German. From Science Abstracts.   11161

Launch vehicle fractionation for communication spacecrafts; C.S.LORENS (Aerospace Corp, Los Angeles, Calif); IEEE—Trans on Aerospace & Electronic Systems v AES-5 n 6 Nov 1969 p 1009-10; Communication satellite models are developed so as to indicate proper number of spacecraft per launch vehicle and proper spacecraft weight fraction which should be assigned to spacecraft housekeeping.   12879

COMPOSITE MATERIALS

See also Domes and Shells—Stresses; Plates—Stresses; Temperature Measurement.

Interaction of elements of fiber-reinforced composite material; E.S.UMANSKII; Poroshkovaya Metallurgiya n 1 Jan 1969 p 101-7. See also English translation in Soviet Powder Met & Metal Cer n 1 Jan 1969 p 80-5; Uniaxial elongation of composite material reinforced with unidirectional fibers is considered; problem is reduced to investigation of state of stress of cylindrical structure (fibermatrix), loaded at ends with uniform normal stress, applied to matrix; formulas derived permit evaluation of effect of some factors on nature of interaction between fibers and matrix with elongation of sample of unidirectional material cut along fibers, in presence of friction forces only. In Russian.   09119

Construction of theory of stability of unidirectional fibrous materials; A.N.GUZ (Inst of Mechanics, Academy of Sciences, Kiev, Soviet Union); Prikladnaya Mekhanika v 5 n 2 Feb 1969 p 62-70; Material considered consists of isotropic binder and isotropic filler having different rigidities; filler consists of straight fibers

ELECTRIC CONTACTS

Study of contact surfaces with scanning electron microscope; O. JOHARI (IIT Research Inst, Chicago, Ill), F.C.HOLTZ, I.CORVIN; Illinois Inst of Technology—Proc of Holm Seminar on Electric Contact Phenomena Nov 11-15 1968 p 207-12; Scanning electron microscope—with very large depth of focus (about 300 times that of optical microscope), ease of directly examining surfaces, and wide range of magnifications (from 14 to 50,000 X)—offers unique tool for study of electrical contact surfaces of all kinds; studies of growth of surface films in arcing copper contacts showed presence of fused oxide film containing pores, regions of molten and frozen copper, and some cracks; observations made at several stages during growth of this surface layer as function of number of cycles are discussed. 13385

Corrosion

Corrosion through porous gold plate; S.J.KRUMBEIN (Burndy Corp, Norwalk, Conn); Illinois Inst of Technology—Proc of Holm Seminar on Electric Contact Phenomena Nov 11-15 1968 p 67-83; Corrosion of gold plated nickel in sulfur dioxide-containing environments is considered, and morphology of corrosion deposits is related to contact behavior; in contrast to spreading of thin sulfide films over gold plated silver, galvanically produced nickel salts were present mainly as discrete mounds, localized at pore sites; study shows importance of humidity to corrosion; this suggests methods for combating pore corrosion, which involve use of hydrophobic films on contact; examples are recently developed inhibitor-lubricant coatings. 11708

Liquid

Contact resistance between solid electrodes and liquid metal in its dynamic state; R.S.RAMSHAW; Illinois Inst of Technology—Proc of Holm Seminar on Electric Contact Phenomena Nov 11-15 1968 p 165-9; Some qualitative results have been obtained for variation of contact resistance between mild steel and mercury, where independent parameters were area of contact and speed of flow of mercury; in general, for given area of contact, resistance decreases with increase in speed of fluid; what is surprising is that, for range covered and for constant fluid speed, contact resistance increases with increase in contact area; current density had no significant effect. 10167

"Filling up" sliding electrical contacts for homopolar engine; J. P.CHABRERIE (Maitre de Conferences et Faculte des Sciences de Paris, France), A.MAILFERT, J.ROBERT; Illinois Inst of Technology—Proc of Holm Seminar on Electric Contact Phenomena Nov 11-15 1968 p 157-64; Use of liquid metal or alloy for electrical contact depends upon choice of pure metal or definite compound and upon suppression of any gas-metal interface; for homopolar engines "filling up" of gap with liquid metal is advantageous solution; theoretical study of different types of losses related to this technique is presented; it shows that in some cases whole liquid ring turns with rotor avoiding thus any short circuit; necessary conditions which allow good efficiency are indicated; experimental devices are described making clear interesting features of this type of contact. 11019

ELECTRIC CONTROL

See also Automatic Control; Electric Drive; Electric Generators —Control; Electric Motors—Control; Electric Transformers; Radio Modulators.

Prevention and treatment of noise in control signals; F.G.WILLARD (Westinghouse Electric Corp, Pittsburgh, Pa); IEEE—Trans on Industry & Gen Applications v IGA-5 n 3 May-June 1969 p 266-72; Signal recovery and noise rejection are accomplished by invoking principle of contrast, discriminating between useful information and noise as result of some characteristic difference between them; various practical means to effect and employ such contrast are examined, and number of selected examples are presented; role of proper wiring layout and grounding is stressed; high level noise can do more than obscure information in signal; number of instances exist in which actual equipment damage has been caused by noise. Paper 68TP 119-IGA. 12901

ELECTRIC CONVERTERS

Frequency Converters

Gate control circuit for thyristor cycloconverter; S.MIYAIRI (Tokyo Inst of Tech, Japan), M.SHIOYA; Elec Eng in Japan (English translation of Denki Gakkai Zasshi) v 88 n 12 Dec 1968 p 31-41; Operating characteristics of gate control circuit converters used widely as power sources for speed control of a-c motors; relation between input control voltage and gate firing angle is expressed in terms of various parameters associated with cycloconverter performance. 14 refs. 11727

Frequency Multipiers

Interference factors in frequency multipliers with diodes, (Stoererscheinungen in Frequenzvervielfachern mit Dioden); P.BOBISCH, C.SONDHAUSS; Int Elektronische Rundschau v 23 n 10 Oct 1969 p 256-8; Investigated are the most frequent causes of noise disturbing the operation of diode-type frequency multipliers, e.g., interfering oscillations, hysteresis, and start delays. In respect of these interference factors, step-recovery diodes are shown to cause less difficulties than varactors. In German. From Science Abstracts. 11192

Valve type transformerless frequency multipliers with direct coupling; G.V.CHALYI, O.A.MAEVSKII; Elektrichestvo n 7 July 1969 p 31-6; High efficiency multiplier with envelope curve is examined, having very simple control system and sufficiently rigorous output characteristics, as well as load symmetry when fed from three phase networks; oscillograms of output voltages and currents of thyratron multipliers. In Russian. 13361

ELECTRIC DISCHARGE

See also Accelerators—Cyclotron; Electric Circuits—Switching; Electric Contacts; Electron Tubes—Magnetron; Lasers—Noise; Welding, Electric Arc.

Diagnostics on steady-state cross-flow arcs; D.M.BENENSON (State Univ of New York, Buffalo), A.J.BAKER, A.A.CENKNER,Jr; IEEE—Trans on Power Apparatus & Systems v PAS-88 n 5 May 1969 p 513-21; Novel cross-flow arc system has been developed which permits attainment of repeatable arc configurations, oriented in vertical plane containing horizontal upstream flow vector, symmetrical in cross section, and symmetrical with respect to midlocation between electrodes; bow-shaped arc is found at lower currents; cusp-shaped arc is obtained at higher currents; transition regime is observed at intermediate levels; development of optical system to observe arc simultaneously at many different azimuthal locations. 16 refs. Paper 68 TP 648-PWR. 09469

Effect of prebreakdown stresses on breakdown impulse voltages of electrode arrangements in air, (Einfluss von Vorbeanspruchungen auf die Durchschlag-Stosspannungen von Elektroden-Anordnungen in Luft); W.RASQUIN; ETZ (Ed A) v 90 n 17 Aug 15 1969 p 415-20;

Plate-plate, point-plate and coaxial cylinder arrangements were investigated using d-c voltages and, in additional test series, prestressed with a-c voltages with frequency of 50 Hz; impulse voltages were superposed on these stresses after long stressing period; magnitude of relative breakdown frequency in case of widely varying composition of mixed voltages; surprising effects were found of even comparatively small prestresses in range of switching voltages, particularly with coaxial cylinder arrangement. In German.
                                                                    12692

Bipolar space charge limited current between coaxial cylinders and concentric spheres; H.AMEMIYA (Inst Phys & Chem Research Tokyo, Japan)—Sci Papers v 63 n 1 Mar 1969 p 1-6; Equations are solved numerically on assumption that initial velocity for different values of space charge neutralization factor is zero; results give exact rate of increase in electron current as compared with unipolar current; inside cathode is shown to be favorable as it always gives larger current than that of plane electrodes and easily temperature limited current; this effect may be due to concentration of positive ions near cathode; thus, two-dimensional inside cathode structure is recommended for thermionic converter with positive ion injection.   12957

Plasma. See also Electric Lamps—Discharge; Helium; Light—High Intensity.

Investigation of electric discharge through laser produced spark; V.I.VLADIMIROV (Physicotechnical Inst im. A.F.IOFFE, Academy of Sciences, Leningrad), G.M.MALYSHEV, G.T.RAZDOBARIN, V. V.SEMENOV; Zhurnal Tekhnicheskoi Fiziki v 39 n 5 May 1969 p 906-10; Laser generated spark between two electrodes produces behind shock wave hot zone which creates conditions for electric discharge; separation of shock wave front from boundary of hot zone occurs 100 nsec after beginning of spark when shock wave velocity is 3×10^5 cm/sec; beginning of discharge depends on location of electrodes with respect to center of hot zone; produced discharge is self-sustained. 4 refs. In Russian.   09712

Electric characteristics of pulsed discharge in liquid—1, 2; I. Z.OKUN; Zhurnal Tekhnicheskoi Fiziki v 39 n 5 May 1969 p 837-61; Approximate criterion for similarity solution in water was obtained; low temperature (ten thousands of degrees), dense, and high (ten thousands of atmosphere) pressure plasma is obtained; periodic regime of discharge is studied; in Pt 2, aperiodic discharge in water and pulsed discharge in transformer oil are investigated; formulas for determination of pressure and resistivity during first half period of discharge in water are presented. 13 refs. In Russian.   09713

Spectroscopic investigation of low voltage arc in cesium at low pressure; M.A.LEBEDEV, S.A.MAEV, G.G.MOROZOVA, I.I. SMIRENKINA; Zhurnal Tekhnicheskoi Fiziki v 39 n 4 Apr 1969 p 673-7; Results of measurements of electron density and temperature at excitation of some lines of scattering sequence in low voltage cesium arc at low pressure are presented; it is found that plasma is strongly unstable; attempt is made to explain results obtained on basis of low voltage arc theory. 9 refs. In Russian.   09999

Effect of traces of oxygen on nitrogen discharges and afterglows; B.BROCKLEHURST (Univ of Sheffield, England), R.M. DUCKWORTH; Phys Soc—Proc (Atomic & Molecular Physics) Ser

ELECTRIC DISCHARGE

2 v 1 n 5 Sept 1968 p 990-6; Intensity of first negative bands in microwave discharge in flowing nitrogen is reduced much more than those of other systems by traces of oxygen—this is ascribed to removal of N_2^+ (singly ionized nitrogen); experiments on afterglows are interpreted in terms of recombination of N atoms catalyzed by N_2^+ in pink afterglow. 10698

ELECTRIC DISTRIBUTION

See also Electric Capacitors; Electric Lines; Electric Transformers; Electric Transmission—Stability.

34.5-Kv distribution. 21-Year solution to voltage regulation; D. L.ANDREWS; Transmission & Distribution v 21 n 10 Oct 1969 p 119-22; Favorable experience with voltage regulation on 34.5 kv distribution lines supplying heavy irrigation-pump loads; only two feeders in all of current twenty-six 34.5-kv circuits supplied by 12 substations have required step-type voltage regulators out on line in addition to usual equipment at substation, and these were for unusually heavy loads far out on feeders. 10924

Statistical and probability models of loads of urban electric power networks; Yu.A.FOKIN (Moscow Energetic Inst); Izvestiya Vysshikh Uchebnykh Zavedenii, Energetika n 8 Aug 1969 p 16-21; The theory of random functions is applied to load analysis, and correlation functions are given for typical consumers. It is shown to be expedient, for simplifying calculations, to replace random functions by random values in the cases of stationary and ergodic intervals. In Russian. 11536

Distribution at 34.5 KV proves economical and practical; C.L. RUDASILL; Pennsylvania Elec Assn—Eng Sec for meeting May 13-14 1969 Appendix F 5 p; Experience with 34.5 kv distribution by Virginia Electric and Power Co. Distribution at 34.5 kv has been found to be the economic solution to distribution problems in all areas of the system—in rapidly growing high load density urban areas, where there is no space to build the additional substations and circuits required by lower voltages, and also in rural areas, where the higher voltage removes voltage drop limitations on long lines, thus avoiding or postponing new transmission construction. 12867

High voltage distribution progress, procedures and planning at Pennsylvania Electric Co; R.D.GOOD; Pennsylvania Elec Assn— Eng Sec for meeting May 13-14 1969 Appendix E 7 p, 10 plates; The distribution of utility combining highly sophisticated network supply of the metropolitan areas, to the well developed and heavily loaded urban areas, and to the widely scattered customers supplied over long, lightly loaded rural lines. 12869

Moon. See Moon—Bases.

Reliability. See Electric Generators.

Underground. See Electric Transformers—Protection; Electric Transformers—Temperature.

ELECTRIC DRIVE

See also Automatic Control; Electric Motors—Braking; Electric Motors—Induction; Integrated Circuits; Mine Hoists—Control; Paper Machinery—Electric Drive; Rolling Mills—Electric Drive; Servomechanisms—Servomotors.

Types of D-C static drives and their application; C.E.GREEN (St. Regis Paper Co, Jacksonville, Fla); IEEE—15th Annual Pulp & Paper Industry Tech Conference, Atlanta, Ga, May 7-9 1969 paper 7, 12 p; Direct-current static drive is electrically regulated, continuously adjustable speed drive, utilizing silicon controlled rectifiers to obtain adjustable d-c voltages for operation d-c motor from a-c line supply; principal elements are reference supply, regulator amplifier, SCR firing circuit, SCR conversion unit, d-c motor, and magnetic control devices; topics covered include principal elements, specifications, characteristics and applications.

09084

Estimation of dynamics of certain d-c reversive drives; V.T. BARDACHEVSKII (L'vov Polytechnical Inst, Soviet Union), R.S. KISHKO; Elektrichestvo n 7 July 1969 p 27-31; Analysis of drive systems consisting of motors fed by controlled rectifiers with feedback; method is described based on determination of interrelation between dynamics of motor starting and braking currents in current limitation sections at equal motor speed; numerical results are compared with actual laboratory oscillograms of generator motor systems with thyristor excitation. In Russian.   11706

Dynamic behavior of d-c control drives fed from thyristors, (Zum dynamischen Verhalten thyristorgespeister Gleichstrom-Regelantriebe); K.FIEGER; ETZ (Ed A) v 90 n 13 June 20 1969 p 311-16; It is shown that, if integrated controllers are used, and if there is adequate smoothing of actual value, positioning unit can be regarded as inertialess proportional action element; consequences which result for setting of controllers are demonstrated by giving example of current-speed control of d-c machine via thyristors in parallel push-pull circuit free from circular currents. In German.   11724

Multimotor drive for coating machine using synthetic resin; F.BRUNNER; Bul Oerlikon n 387-388 Apr 1969 p 37-9; Drive of PVC coating machine is used to coat different materials such as paper, textiles and floor covering with layer of PVC; PVC granulite is melted and coated on base material; drive system consists of four speed controlled d-c motors; heated melting rollers are driven by motors 1, 2 and 3 through reduction gears; motor 4 drives dosing plant; motor 2 serves as guide motor; motors 1 and 3 can be set with relative speed deviation of 0 minus 15% and 0 plus 15% respectively, with respect to guide motor; this adjustment is necessary so that PVC web can be taken up by different rollers.

12323

Variable Speed. See Electric Motors—Induction.

ELECTRIC EQUIPMENT

New results concerning irradiation on electronic and electrotechnical equipment, (Resultate noi privind actiunea radiatiilor asupra materialelor si dispozitivelor electronice si electrotehnice); G.IONESCU (Institutul Politehnic, Bucharest, Roumania), V.FO-CHIANU; Electrotehnica v 17 n 6 June 1969 p 232-40; Effects of radioactive radiation are reviewed especially on switchgear insulation. In Roumanian.   10402

Explosionproof

Probability factors as guide for area classification and selection of electrical equipment; J.M.BENJAMINSEN (N.V.Nederlandse Staatsmijnen/DSM, Geleen, Netherlands), P.H.VanWIECHEN; IEEE —Trans on Industry & Gen Applications v IGA-5 n 3 May-June 1969 p 242-9; Mathematical model is given that shows how mean time between explosions, which is measure of safety level, can be derived from frequency in presence of explosive gas mixtures, mean time between failures of electric equipment, and inspection interval; results indicate that area classification, which does not make allowance for inspection interval and mean time between failures of electric equipment is inappropriate. Paper 68 TP 114-IGA.   13042

Grounding

Contact voltage and potential of complex grounding devices in homogeneous ground; A.A.VORONINA; Elektrichestvo n 7 July 1969 p 52-6; Electrolytic tank study of grounding device geometry, such as appropriate mesh configuration, length and density of vertical electrodes and effect of deepness of grounding; simple formulas and their numerical coefficients are derived for engineering calculation of optimal grounding devices. In Russian.
10191

Economic design of grounding systems—1, (Zur wirtschaftlichen Bemessung von Erdungsanlagen); M.JURKE; Elektrie v 23 n 5 May 1969 p 210-12; Technical-economic criteria are determined for vertical arrangements of grounding plates. The relation between the distance/length ratio, the number of plates, ground resistance required, and specific costs are presented. In German.
12891

Economic calculations of grounding equipment; M.YURKE; Elektrichestvo n 7 July 1969 p 77-9; Method for selecting optimum dimensions for simple and complex grounding systems consisting of vertical electrodes and connecting strips. In Russian.
13433

Insulation. See Electric Coils—Manufacture.

Maintenance

Optical oscillograph speeds troubleshooting; R.G.SMITH (Federal Dept of Transport, Ontario, Canada); Elec Construction & Maintenance v 68 n 10 Oct 1969 p 133-6; Use of high-speed, direct-writing oscillographic recorders to pinpoint trouble sources often overlooked by other instruments in determining starting current of motors or to spot short-duration transient voltages that sometimes disrupt operation of computers and data processing equipment.
11906

Materials

Distribution of statistical zones in tension phase range of cutting edges running parallel with direction of rolling in demagnetized Goss texture, (Die Verteilung statistischer Domaenen im Bereich der Zugphase von parallel zur Walzrichtung verlaufenden Schnittkanten bei entmagnetisierter Gosstextur); G.M.FASCHING, H.HOFMANN, M.SCHUECKHER; Elin-Zeit v 21 n 1-2 June 1969 p 39-41; Stress effect is analyzed using powder patterns of statistical domain zones. 14 refs. In German.
12945

Molded. See Electric Cables—Connectors.

Power Supply. See Moon—Bases.

Printed

Frame design and tension control for precise screen process printing of micro-circuits with mesh screens and etched masks; H.L.CORONIS (Industrial Reproductions, Inc, Nashua, NH); Int Soc for Hybrid Microelectronics—Proc 1st Tech Thick-Film Symposium, Palto Alto, Los Angeles, Calif, Feb 21-4 1967 (recd 7/12/68) p 21-34; Increasing demands for tighter printing controls in field of micro-circuits have resulted in new frame design; among factors considered in preparing design were flatness, rigidity, and provision for positioning in printing machine; effect of tension control for precise screen process printing of micro-circuits is included; value of machined frames and close observance of screen

INTEGRATED CIRCUITS

Elec Eng Conf Publ No. 49, Stevenage, Herts, 1968 p 70-80; Consideration is given to factors affecting development of microcircuits; current state of art is discussed and some factors affecting future trends are examined including development of new techniques and falling cost per circuit unit; finally some examples of integrated circuit systems are discussed giving consideration to probable configuration of next generation designs; examples chosen are heat control, speed controlled reversing drive, and machine tool numerical control system.   12762

Manufacture. See also Computers—Circuits.

Tooling and part handling problems in production of thick film microcircuits; D.C.HUGHES,Jr (Precision Systems, Co, Inc, Bound Brook, NJ); Int Soc for Hybrid Microelectronics—Proc 1st Tech Thick-Film Symposium, Palo Alto Los Angeles, Calif, Feb 21-4 1967 (recd 7/12/68) p 102-12; Variety of part holding tools of fixtures used at screen print or adjust phases of production process in flat bed or rotary dial machines are discussed; types covered range from single locating pin fixtures to elaborate clamping vacuum fixtures and compares requirements for substrates of different materials in plate and disk form; various ways to feed and transfer parts between various process operations are examined; various process operations from raw substrates through clean, print, dry, fire, lead-attach, adjust, hybrid component mounting, encapsulation and test are discussed.   09490

Aspects of reliability in monolithic integrated circuit device interconnections; C.W.PITT; Microelectronics v 2 n 5 May 1969 p 38-42; Problems associated with interconnection and packaging systems used in ICs are discussed, particularly from viewpoint of reliability; pertinent guidelines to serve engineer as specification aid are presented which are backed up with tabulated results of tests on wire bond strength.   10297

Technology of multilayered thick film circuit arrays; R.G. LOASBY; Microelectronics v 2 n 4 Apr 1969 p 12-16; Some of problems that arose while developing thick-film multilayered circuit array are discussed; process is described which was developed to ensure compatibility with existing methods of device mounting and thin-film processing; number of basic design concepts are introduced to produce high component packing density and to maintain this density through multiple layers of circuitry.
10303

Thermal considerations for systems using hybrid ICs; R.C.CHU; Microelectronics v 2 n 7 July 1969 p 32-7; As packaging and component densities increase, so power dissipation and thermal considerations become real problems; three main problem areas are discussed, namely, heat conduction from chip through package, heat removal from surface of package, and maintaining temperature of cooling system; conduction from chip to package depends mainly on chip attachment process used, whereas heat removal from package depends on package configuration and cooling conditions; trends in liquid cooling systems are projected for future requirements. 14 refs.   10602

Complex monolithic arrays. Some aspects of design and fabrication; J.A.NARUD (Motorola Inc, Phoenix, Ariz), C.D.PHILLIPS, W.C.SEELBACH; Microelectronics v 2 n 7 July 1969 p 12-17; Concept of large-scale integration (LSI) is discussed; it is shown what is practical with present techniques, and what is involved in terms of circuit performance, IC fabrication, packaging and testing, and what direction future development is likely to take.   10604

Future role of film circuits in microelectronics; P.L.KIRBY (Welwyn Electric Ltd, Northumberland, England); Microelectronics v 2 n 7 July 1969 p 18-22; Development of both thick and thin-film techniques is discussed, along with their relative merits; growing areas of application for which they are ideally suited are indicated.
10605

Improved thin-film conductors for microcircuits; P.S.KENRICK; Microelectronics v 2 n 6 June 1969 p 32-5; As integrated circuits (IC) become increasingly complex it is necessary to make use of several layers of metal interconnections on chip; if ICs using such multilayer systems are to be as reliable as conventional devices, then metal film used must combine good substrate adhesion properties with surface free from pinholes and other irregularities, together with low electrical resistance; in order to develop technique for production of such films, investigation was undertaken and results are presented.
10606

Impact of silane epitaxy on IC fabrication; G.SHRANK, E.De BENE-DITTI; Microelectronics v 2 n 6 June 1969 p 24-7; Technique for growth of epitaxial layers on silicon using silane instead of silicon tetrachloride is described; it eliminates problems of out-diffusion, autodoping and buried layer shift; experimental results are given for various conditions of growth, together with characteristics of growth layers.
10607

Ultra-microminiaturization precision photography for electronic circuitry; C.R.HANCE ed; Soc of Photographic Scientists & Engrs, Washington, DC, 1968, 210 p; Proceedings include 16 papers which review the process in detail and explain the role that photography plays in the fabrication of microminiature electronic components; major topics covered include equipment design and performance, optics for silver halide exposure, optics for photoresist exposure, photosensitive materials, techniques and processes, metrology, inspection, semiconductor fabrication, unconventional exposure methods, and resolution and quality evaluation. Following papers presented—First reduction camera; H.L.COOPERMAN; Photo mask making equipment; P.L.OSTAPKOVICH; Precision plate camera for research in micro-photography; J.H.ALTMAN, H.C.SCHMITT, Jr; Optical systems for direct projection on photoresists; M.J.BUZAWA, G.G.MILNE, A.M.SMITH; Series of high performance reduction lenses for production of microelectronics; R.E.TIBBETTS, J.S.WILCZYNSKI;Cleaning and processing of high resolution plates, and techniques for evaluating their quality for critical applications; J.LEVINE; Micro-miniature glass plate processing; O.FISHER; Micro photo masking making—present and future; S.FOK; Photographic image structure evaluation; C.S. McCAMY; Developments in Japan in commercial optics; S.HELD; Ultra-micro demands on photoresist; L.E.MARTINSON; Sharpness of photographic plates for microelectronics; M.DeBELDER, H. PHILIPPAERTS, R.DUVILLE, D.SCHULTZE; Unconventional methods of photoresist exposure; J.DEY, S.HARRELL; Equipment for projection exposure of photoresist; H.B.LOVERING; Developments in Japan in commercial optics; S.HELD; Mask inspection techniques; G.M.HENRIKSEN.
10893

IEEE—Wescon Technical Papers Aug 19-22 1969, v 13 Pt 1. Manufacturing Technology; Anon; Wescon, Los Angeles, Calif, 1969, various pagings; Collection of 16 papers concerned with handling microcircuits automatically, use of computers as aid in manufacturing control, automatic production of semiconductors, and computer-aided testing, management and implementation. 15 papers were abstracted separately.
12027

Reliability. See Integrated Circuits—Testing.

Testing

Microcircuit testing. Matching value with cost; J.P.BRAUER; Microelectronics v 2 n 6, 7 June 1969 p 17-23, July p 23-31; Experience gained from severe reliability proving program of U S Air Force is reported; use of integrated circuits (IC) and their cost; cost of testing, failure modes and mechanisms, and failure rate predictions; assessment of IC reliability and testing; results of tests to determine amount of testing required to verify reliability of family of transistor-transistor logic (TTL) devices for application in various projects. 35 refs.   10409

Computer-controlled on-line testing and inspection; P.H.GOEBEL (General Radio Co, W.Concord, Mass); IEEE—Wescon Tech Papers v 13 pt 1 Session 8/3 1969, 5 p; Problems associated with testing integrated circuits are discussed; it is shown how test fixture design and testing of even most complex modules was made routine task at given electronics plant; it is concluded that computer-controlled test systems used by manufacturing facilities have necessary future dictated by ever-improving technology.   12340

INTERFEROMETERS

See also Heat Transfer—Measurement; Radiometers.

Accessories

Interference grating, (Das Interferenzgitter); M.RICHARTZ; Optik v 29 n 2 May 1969 p 146-9; New kind of grating, called "interference grating" has been introduced into interferometry; plane parallel glass plate provided with transparent and silvered stripes alternatively arranged, is considered multiple Lloyd's mirror; practical applications of such grating in interferometers to replace semireflecting devices are discussed. In German.   10215

INTERMETALLIC COMPOUNDS. See Iron—Intermetallic Compounds.

INTERNAL COMBUSTION ENGINES

See also Gasoline; Oil Well Production—Flooding.

Combustion

Computer model for prediction of combustion chamber geometrical characteristics; K.D.S.R.SOMAYAJULU (Indian Institute of Tech, Kharagpur, India), J.K.SUBRAHMANYAM; ASME—Paper 69-WA/DGP-6 for meeting Nov 16-20 1969, 8 p; Volume of charge burnt and surface area of propagating flame influence rate of pressure rise in spark ignition engine and, although combustion chamber geometry and spark plug location are important factor in design, there is not rational method of predicting combustion chamber characteristics; method that envisages determination of volume common to sphere and combustion chamber; limitations of method and steps to obviate them are presented.   09793

Costs. See Internal Combustion Engines—Design.

Deposits. See Lubricating Oil—Testing.

Design

Internal combustion engine. New roles for old performer; G. W.NEAL; Consulting Engr v 33 n 1 July 1969 p 88-94; Mathematical analysis of relevant data for accurate prediction of relative costs based on total energy concept, under arrangement

INTERNAL COMBUSTION ENGINES

of which engines are used both as source of work for electric generation and as source of heat; how to apply discounted cash flow analysis to determine which type of power will do least expensive work; prime mover heat recovery capabilities.   11843

Exhaust Gases. See also Fuels—Research.

Effects of air-fuel ratio on composition of hydrocarbon exhaust from isooctane, diisobutylene, toluene, and toluene-n-heptane mixture; J.S.NINOMIYA, A.GOLOVOY; SAE—Paper 690504 for meeting May 19-23 1959, 11 p; Fuels listed in title are burned in single cylinder engine and produce following results; olefin concentrations in exhaust, in general, are temperature dependent and are maximum near equivalence ratio of 1.0; high concentration of acetylenes from diisobutylene and toluene fuels suggests that engine combustion processes are largely pyrolytic in nature; addition of 25.% n-heptane to toluene increases concentrations of those exhaust products formed via alkylation; combustion of diisobutylene, olefinic fuel, above equivalence ratio of 1.0 produces less olefinic materials than isooctane fuel; thus total photochemical reactivity of diisobutylene exhaust is less than that of isooctane above this equivalence ratio. 48 refs.   09805

Vibrations

Study on small-sized torsional damper for high speed internal combustion engines; M.ISHIZUKA (Tokyo Machinery Works, Japan); Mitsubishi Heavy Industries, Ltd—Tech Rev v 6 n 2 May 1969 p 122-30; There is growing demand for small-sized torsional damper with big damping force for use in high speed internal combustion engines; application of theory of two-mass system is most desirable, but it requires further analysis to adopt it in internal combustion engines of multimass system; author analyzes theory of two-mass system as applied to general conditions, and introduces small-sized damper developed for practical use on basis of results of fundamental test on rubber and silicone fluid which are adopted for economical advantages.   10967

INTERPLANETARY COMMUNICATION. See Radio Communication—Interplanetary.

INVENTORY CONTROL

Bills of material data structures for use in computerized inventory planning; F.H.VEITH,Jr; SAE—Paper 690486 for meeting May 19-23 1969, 8 p; Advanced techniques to structure bills of material file on random access devices so that they may be used effectively by computer are demonstrated; balancing of requirements against supply orders for each material item in product structure produces economical inventory plan and can be achieved in integrated data store bills of material file; basic data file is structured in manner called "integrated data store", or "ring structure", which permits group of related records to be tied together; tie is accomplished by storing, within each record of group, address of next record.   11883

IODINE COMPOUNDS. See Crystals—Optical Properties.

ION BEAMS. See Accelerators—Ion Sources; Plasmas—Waves.

ION EXCHANGERS

See also Adsorption; Feedwater Treatment—Ion Exchangers; Plastics—Polyester; Sand, Foundry—Binders.

Infrared spectrum studies of phosphorus-containing ion exchangers; B.N.LASKORIN, L.A.FEDOROVA, I.A.LOGVINENKO, N.P. STUPIN; Zhurnal Prikladnoi Khimii v 42 n 3 Mar 1969 p 522-9. See also English translation in J Applied Chem of USSR v 42 n 3 Mar 1969 p 490-5; Phosphorus-containing cation exchangers, which are active complexing agents for many metal cations, are widely used for chromatographic separation of metals; study of IR spectra of cation exchangers made from phosphonic acids and arylalkylphosphonic and acrylic acid copolymers showed differences in vibration frequency of PO group and in band intensity and halfwidth; hydrogen bonds were detected between -$PO(HO)_2$ groups in arylalkylphosphonic acid copolymers; hydrogen bonds of mixed type are formed between -$PO(OH)_2$ and -COOH groups in polyfunctional carboxylic-phosphonic ion exchangers; influence of effect on sorption properties of carboxylic-phosphonic cation exchangers. 20 refs. In Russian. 09529

Mechanism of trace counterion transport through ion-exchange membranes; W.J.BLAEDEL (Univ of Wisconsin, Madison), T.J. HAUPERT, M.A.EVENSON; Analytical Chem v 41 n 4 Apr 1969 p 583-90; Equation for flux of trace counterion between two solutions separated by ion-exchange membrane is derived and experimentally verified; equation is tested using both synthetic anion- and cation-exchange membranes for systems containing singly and multiply charged trace and bulk ions; experimental evidence is also presented to indicate influence of surface exchange reaction on overall rate of ion transport. 10819

Resin phase concentration dependence of cation exchange equilibria; M.M.REDDY (State Univ of New York, Buffalo), J.A.MARINSKY; Am Chem Soc—Div Polymer Chem v 10 n 2 Papers for New York meeting Sept 1969 p 1152-6; Dependence of cation exchange equilibria on resin phase concentration was studied in dilute perchloric acid solution of concentration .168, .016 and .001 molal using trace concentrations of Ca^{+2}, Sr^{+2}, Co^{+2}, Ni^{+2}, Zn^{+2} and Cd^{+2}; resin phase concentration range was 1.24 molal to 6.91 molal, being determined by nominal percent divinyl benzene content of polystyrene sulfonate based ion-exchanger. 11837

Preparation of carbocyclic cation-exchange fibers; E.Ya.DANILOV (S.M.Kirov Leningrad Inst of Textile and Light Industries), A.I.MEOS, L.A.VOL'F, Yu.K.KIRILENKO; Zhurnal Prikladnoi Khimii v 42 n 4 Apr 1969 p 968-70. See also English translation in J Applied Chem of USSR v 42 n 4 Apr 1969 p 926-8; Simple method of preparing carbocyclic cation-exchange fibers having high chemical and thermal stability is based on condensation of maleic anhydride with dehydrated PVA fibers. The static cation-exchange capacity of the resulting fibrous cation exchanger reaches 8 meq/g, dependent on the treatment conditions. In tests by the methods for determination of the resistance of ion exchangers to oxidizing agents, alkalies, and acids, their exchange capacity remained unchanged. In Russian. 12605

IONIZATION

See also Gases—Ionization; Metal Corrosion; Microscopes—Electron; Vapors.

Initial ionization rates and collision cross sections in shock-heated argon; T.I.McLAREN (Queen's Univ of Belfast, Northern Ireland), R.M.HOBSON; Physics of Fluids v 11 n 10 Oct 1968 p 2162-72; Investigation is made in low-pressure, 4.5-in.-diam, shock tube using double electrostatic probes to measure ionization rates; experimental conditions cover range of initial pressures of spectroscopically pure argon from 0.2 to 3.0 torr and temperature from 7000 to 12,000 K; maximum impurity level at 1 torr pressure is less than 7 ppm. 22 refs. 11088

IONIZATION CHAMBERS

See also Amplifiers—Direct Coupled.

Chambers for use with semiconductor detectors; Yu.F.RODIO-NOV, I.V.NAUMOV; Instruments & Experimental Techniques (English translation of Pribory i Tekhnika Eksperimenta) n 2 Mar-Apr 1969 p 322-4; Ionization chambers for use in semiconductor α spectrometers and in α-γ-coincidence devices are described. Both are used for isotopic analysis of mixtures of α-emitting elements. Drawings and photographs are included.   12377

IONOSPHERE

See also Computers—Data Processing; Lasers—Liquid; Radar—Measurements; Radio Waves—Reflection; Satellites—Design; Satellites—Instruments; Satellites—Telemetering.

Topside ionosphere during geomagnetic storms; E.S.WARREN (Communications Research Centre, Ottawa, Ont); IEEE—Proc v 57 n 6 June 1969 p 1029-36; State of knowledge of ionospheric storms in topside ionosphere is reviewed; although much remains to be understood concerning ionospheric storms, many of storm-time phenomena can be interpreted in terms of drift motions produced by storm-time currents, and by ambipolar diffusion. 40 refs.
10739

Ionospheric ion composition deduced from VLF observations; R.E.BARRINGTON (Communication Research Centre, Ottawa, Ont); IEEE—Proc v 57 n 6 June 1969 p 1036-41; VLF observations from Alouette satellites have provided information on ion composition of ionosphere; this is derived from two sources, namely, noise bands with lower frequency cutoff at lower hybrid resonance (LHR) frequency, and ion whistlers; LHR frequency depends on harmonic mean mass of ionic constituents; ion whistlers provide good measurements of relative abundances of protons and helium ions, as well as proton density and proton temperatures. 25 refs.
10944

On prediction of F-layer penetration frequencies; L.E.PETRIE (Communications Research Centre, Ottawa, Ont), G.E.K.LOCK-WOOD; IEEE—Proc v 57 n 6 June 1969 p 1025-8; Predictions of F-layer penetration frequencies by CCIR (International Consultative Committee of Radiocommunications) are compared with 1963 Alouette topside-sounder data; data used for these comparisons were collected from 45 to 165 degrees W longitude in northern hemisphere; it was found that CCIR predictions differ by as much as 0.7 MHz from Alouette data.   10945

Frequency shifts observed in Alouette II cyclotron harmonic plasma resonances; R.F.BENSON (NASA, Goddard Space Flight Center, Greenbelt, Md); IEEE—Proc v 57 n 6 June 1969 p 1139-42; Cyclotron harmonic plasma resonances observed by Alouette II were investigated using technique that yielded frequency measurements with accuracy of order of 1 kHz; comparison of earlier Alouette I results with Alouette II results indicates that bulk of original excitation volume associated with higher cyclotron harmonic resonances is confined to region considerably less than one antenna length in radius from antenna, that size of this region decreases with increasing harmonic number n, and that it increases significantly as radiating antenna becomes parallel to direction of earth's magnetic field.   10946

Model studies of kinked Z trace in topside ionograms; L.COLIN (NASA, Ames Research Center, Moffett Field, Calif), K.L.CHAN;

SEMICONDUCTOR DEVICES

cluding p-n junction diodes, Schottky barrier diodes, and interconnected arrays of isolated devices on semi-insulating substrates.
10316

Analysis of transient response of p-n junctions by method of fluxes; I.M.BESKROVNYI; Radiotekhnika i Elektronika v 14 n 4 Apr 1969 p 702-9. See also English translation in Radio Eng & Electronic Physics (pub by IEEE) v 14 n 4 Apr 1969 p 601-7; In analysis of transient performance is proposed to use method of fluxes, that gives solutions valid even to t approaches zero; method of analysis using flow equations is developed that possesses greater accuracy and added simplicity and flexibility. In Russian.
10874

Depletion layer calculations for error function diffused junctions; P.R.WILSON (A.E.I.Semiconductors Ltd, Lincoln, England); Solid-State Electronics v 12 n 4 Apr 1969 p 277-85; Expressions are derived for voltage dependance of depletion layer width, maximum electric field and capacitance of error function diffused plane cylindrical and spherical p-n junctions; these are also shown in graphical form for junctions in silicon for ratios of surface to background concentration covering range 10^2 to 10^7; it is shown that junctions can be considered to be either linearly graded or abrupt with transition range covering about two decades of voltage; results obtained are very similar to those for gaussian diffused junctions.
12064

Heterode strain sensor. Evaporated heterojunction device; R.M. MOORE (RCA Labs, Princeton, NJ), C.J.BUSANOVICH; IEEE—Trans on Electron Devices v ED-16 n 10 Oct 1969 p 850-5; New type of strain sensor has been demonstrated in laboratory; basic device is p-n heterojunction diode that is fabricated by vacuum evaporation techniques directly onto flexible substrate; its two important properties are—its mechanical input characteristics are determined by flexible substrate used in fabricating heterojunction and it functions as low-output-impedance voltage source modulated by substrate surface strain; heterojunction combination that has been most commonly used is CdSe on Se; typical devices exhibit voltage/strain sensitivity of 500 v/unit strain, with differential resistance of 100 ohms, when biased at 5 mA.
12174

Semiconductor mechanical sensors; J.J.WORTMAN (Research Triangle Inst, Research Triangle Park, NC), L.K.MONTEITH; IEEE—Trans on Electron Devices v ED-16 n 10 Oct 1969 p 855-60; Purpose of paper is to briefly review concepts proposed to explain piezojunction effect and to describe techniques employed in its application; several experimental transducers which have been constructed and evaluated are described. 14 refs.
12240

Manufacture

Joining semiconductor devices with ductile pads; L.F.MILLER; Microelectronics v 2 n 4 Apr 1969 p 24-9; Solder reflow of copper-ball devices has been shown to be reliable, effective, and ideal for automation; technique which prevents such solder pads from collapsing and permits large-scale production is described; method involved is termed 'control collapse' and is based on limiting solderable area of module land so that surface tension of molten pad and land solder support the device. 24 refs.
10301

Microwave. See also Semiconductor Devices—Diode; Semiconductor Devices—Oscillator.

Theory for high-efficiency mode of oscillation in avalanche diodes; A.S.CLORFEINE (RCA Labs, Princeton, NJ), R.J.IKOLA,

L.S.NAPOLI; RCA Rev v 30 n 3 Sept 1969 p 397-421; High-efficiency oscillations, previously observed in avalanche diodes and simulated with computer, are explained by means of analytic theory in which all relevant physical processes are clearly displayed and tied together; it is shown how efficiencies of 50 to 60% can be achieved; on basis of theory, design formulas are derived. 12 refs.   09747

Two-dimensional Gunn-domain dynamics; M.SHOJI (Bell Telephone Labs, Inc, Murray Hill, NJ); IEEE—Trans on Electron Devices v ED-16 n 9 Sept 1969 p 748-58; In two-dimensional bulk GaAs devices, each small segment of high-field domain can be considered to move normal to its front, with velocity equal to that of one-dimensional domain having same domain potential; using this simple model, equation describing domain shape in two-dimensional samples was obtained; when edge nucleation effects are taken into account, solution of equation provides good explanation for most of domain motions observed experimentally in various samples of nonuniform shape; experimental observations were made using resistive probe; probe experiments enable one to visualize how domains behave in devices with sudden or gradual changes in width, with sharp or gradual bends, and with multiple terminals. 11 refs.   10727

Semiconductor devices for generation and amplification of microwaves, (Halbleiterbauelemente zur Erzeugung und Verstaerkung von Mikrowellen); W.HEYWANG; Int Elektronische Rundschau v 23 n 10 Oct 1969 p 267-9; The Gunn effect is described. In conclusion, new and still open approaches for RF semiconductor devices are discussed. In German. From Science Abstracts.   11144

Simplified analysis of steadily propagating high-field domain in Gunn effect using piecewise-linear approximation; T.IKOMA (Univ of Tokyo, Japan), H.TORITSUKA, H.YANAI; Electronics & Communications in Japan (English translation of Denshi Tsushin Gakkai Ronbunshi) v 51 n 12 Dec 1968 p 124-32; Properties of highfield dipole domain steadily propagating in n-type GaAs are expressed analytically on basis of three simplified assumptions; these are piecewise-linear approximation of v-E characteristic, field-independent diffusion constant, and triangular field distribution of dipole domain; relations between domain excess voltage and outside electric field are obtained experimentally and compared with theoretical ones. 26 refs.   12653

Oscillator

High-efficiency operation of Gunn oscillator in domain mode; G. S.KINO (Stanford Univ, Calif), I.KURU; IEEE—Trans on Electron Devices v ED-16 n 9 Sept 1969 p 735-48; Examination of problem of obtaining high-efficiency operation of Gunn oscillator is given; results are based on assumed form of I-V characteristic of Gunn diode, ideal voltage and current waveforms being found for high-efficiency operation; theory for square wave of current through and voltage across diode is worked out in detail and shown to predict experimental results well; it is shown that voltage waveform consisting of only fundamental and second harmonic component could give efficiencies in 20 to 28% range, as much as factor of 3 larger than for simple sinusoid; this is reasonable approximation to ideal voltage waveform, which is half sinusoid of voltage. 13 refs. 10452

Photoelectric. See also Semiconductor Devices—Diode.

1-2 Micron (Hg,Cd) Te photodetectors; A.N.KOHN (Honeywell Radiation Center, Lexington, Mass), J.J.SCHLICKMAN; IEEE—Trans on Electron Devices v ED-16 n 10 Oct 1969 p 885-90; Performance of mercury cadmium telluride detectors in 1 to 2 μ

spectral region has been predicted from basic material parameters; photovoltaic devices should be characterized by specific responsivities of 1 A/w for 1000 ohm load when transit time limited to less than 20 nsec; photoconductive detectors made from-n-type material should have radiative lifetimes of 1 msec; feasibility of high performance 1 to 2 μ (Hg,Cd)Te detectors has been demonstrated experimentally; deep junction devices operating at room temperature without bias have been fabricated by impurity in diffusion; detectivities at 1.75 μ approached 10^{10} cm.Hz1/2/w with open-circuit responsivities of approximately 500v/w; in addition 1.5 μ detectors have been fabricated from p-type, 25 ohm cm material. 11 refs.   10895

Temperature cycling effects on solar panels; W.LUFT (TRW Systems, Redondo Beach, Calif), E.E.MAIDEN; IEEE—Trans on Aerospace & Electronic Systems v AES-5 n 6 Nov 1969 p 943-50; Temperature cycling for more than 300 cycles and for temperatures down to -175 C performed on soldered silicon cell assemblies with copper, Kovar, and molybdenum interconnectors showed wide range in failures depending both upon materials used and on interconnector thickness and substrate material.   12002

Switching. See also Instruments—Circuits; Semiconductor Devices —Diode.

Glass switching and amplifying elements, (Schalt- und Verstaerkerelemente aus Glas); A.ENGEL, O.HOLZINGER; Frequenz v 23 n 10 Oct 1969 p 294-301; Novel devices of semiconducting transition-metal oxide glasses are described. These components do not contain any barrier layers and are symmetrical. They can be designed as switching elements or amplifying elements. The layout of these glass elements is described and their electrical characteristics, including dependence on temperature and frequency, are stated in detail. A number of typical applications are also briefly given. In German. From Science Abstracts.   11146

Testing. See Semiconductor Devices—Photoelectric.

Thermal Properties. See Semiconductor Devices—Photoelectric.

SEMICONDUCTORS

See also Crystals—Piezoelectric; Magnetic Materials; Radiation —Detectors; Silica; Vanadium Compounds.

Recombinations via defects in degenerate semiconductors; E.L. HEASELL (Univ of Waterloo, Ont); Solid-State Electronics v 12 n 4 Apr 1969 p 225-8; It is shown that in degenerate materials, for wide class of trap recombination models, it is possible to obtain relevant capture rate equations by using modified definition of Shockley and Read trap parameters; recombination rate expressions, in terms of quasi-Fermi levels may be obtained by using solutions that are already in literature for nondegenerate recombination models. 15 refs.   10320

Computer solutions of Lorentz-field amplification in bounded semiconductor medium; J.R.GOLDEN (RCA Labs, Princeton, NJ), K.K.N.CHANG; J Applied Physics v 40 n 4 Mar 15 1969 p 1936-40; Lorentz field is generated when moving free carriers in semiconductors interact with crossed magnetic field; set of computer results show strong interaction with high-order transverse modes in parallel plate wave guide; as result of this interaction, unstable solutions which lead to generation of growing waves have been found.   10336

SEMICONDUCTORS

Gas adsorption and x-rays studies of internal mated splits in Ge and Si; D.HANEMAN (Univ of NSW., Kensington, Australia), J.T.P.GRANT, R.U.KHOKHAR; Surface Science v 13 n 1 Jan 1969 p 119-29; Electrical properties of small internal splits created in high vacuum in Ge and Si, studied as function of exposure to oxygen and air; recontact of surfaces of split is sufficiently close to exclude gas entry over region which is proportionately larger, smaller split; x-ray examination by Lang technique of splits created in Si in air reveals that observable contrast can disappear in base region of split on recontact; accumulated evidence from electrical, gas and x-ray studies is interpreted to show that splitting technique used permits surface recontact, sufficiently good in base region of split, for lattice restoration to be possible. 15 refs.   10418

Properties of semiconductors useful for sensors; D.LONG (Honeywell Inc, Hopkins, Minn); IEEE—Trans on Electron Devices v ED-16 n 10 Oct 1969 p 836-9; Different types of solid state sensors demand different basic properties of semiconductor materials from which they are made; best transistor material is not necessarily best material for certain sensor; one must choose semiconductor for particular sensor on basis of its fundamental properties, such as energy band structure; for example, piezoresistance sensors are made of silicon or germanium mainly because large effect occurs only in semiconductors, such as these, having complex band edge structures; on other hand, intrinsic infrared photon detector requires energy gap corresponding to longest wavelength to be detected; as well as direct gap, compatibility with integrated circuitry may sometimes be consideration also. 10 refs.   10896

Band structure of magnetic semiconductors; D.ADLER (Massachusetts Inst of Tech, Cambridge), J.FEINLEIB; J Applied Physics v 40 n 3 Mar 1 1969 p 1586-8; Method is proposed for representing effective one-electron density-of-states of materials in which some of valence electrons are extremely localized; if some additional rules regarding nonconducting states and occupation-number dependent states are imposed, this scheme can be used to interpret both electrical and optical properties of magnetic semiconductors; materials NiO and CoO are discussed in detail.   12203

Absence of antiferromagnetism in Ti_2O_3; R.M.MOON (Oak Ridge National Lab, Tenn), T.RISTE, W.C.KOEHLER, S.C.ABRAHAMS; J Applied Physics v 40 n 3 Mar 1 1969 p 1445-7; Possible existence of antiferromagnetism in Ti_2O_3 has important theoretical implications associated with explanation of metal-to-semiconductor transition in this material; Ti_2O_3 has been examined using new technique of neutron-polarization analysis; it is possible to detect small magnetic contributions even when magnetic and nuclear peaks have same scattering vector; at room temperature, no evidence is found for either $\alpha-Fe_2O_3$ or Cr_2O_3 magnetic structure.   12585

Work function and structural studies of alkali-covered semiconductors; R.E.WEBER (Univ of Minnesota, Minn); Surface Science v 14 n 1 Mar 1969 p 13-38; Work function and LEED measurements have been made on clean and alkali-covered Ge and Si surfaces; aluminosilicate ion sources have been used to deposit accurately measurable quantities of Na, K and Cs ions on surfaces normal to Ge(111), Ge(100), Si(111) and Si(100) crystal directions; resulting work function vs alkali metal coverage curves, when plotted on semi-log paper, consist of nearly-straight line segments with distinct breaks; coverages at which breaks occur are consistent with clean surface structures. 29 refs.   12801

Pseudopotential impurity theory and covalent bonding in semi-conductors; M.JAROS (Czechoslovakian Academy of Sciences, Prague), P.KOSTECKY; J Physics & Chem Solids v 30 n 3 Mar 1969 p 497-502; Abarenkov-Heine model potential method is used; method promises quantitative estimation of charge redistribution and polarization effects due to impurity potential; special attention is paid to impurity problem in zero-gap semiconductor; local part of impurity potential is given for antimony atom as impurity in gray tin crystal. 18 refs.   13165

Acoustic Effects. See also Semiconductors—Piezoelectric.

Acoustoelectric amplification in InSb; R.K.ROUTE (Stanford Univ, Calif), G.S.KINO; IBM J Research & Development v 13 n 5 Sept 1969 p 507-9; In study of two modes of acoustoelectric domain oscillation which occur in InSb in transverse magnetic field, lithium columbate transducers on acoustic amplifier were used to measure linear acoustic gain as function of electric and magnetic field and frequency; at low magnetic fields, wavelength of sound waves is less than mean free path of electrons, and macroscopic theories break down; extending microscopic theory of magnetacoustic interactions to include electron drift gives excellent agreement between theory and experiment and results account for two modes of acoustic domain formation.   12799

Bismuth Compounds

Electromechanical properties of bismuth germanate $Bi_4(GeO_4)_3$; H.SCHWEPPE (Philips Zentrallaboratorium GmbH, Aachen, West Germany); IEEE—Trans on Sonics & Ultrasonics v SU-16 n 4 Oct 1969 p 219; As is known from literature, single crystals of bismuth germanate have been grown, and electro-optical properties have been investigated; this correspondence describes electromechanical properties, i.e., elastic, dielectric, and piezoelectric behavior.   10465

Cadmium Compounds. See Semiconductors—Films; Semiconductors—Thermal Properties.

Cadmium Selenide

Current saturation and oscillation in CdSe; S.YEE (Univ of Washington, Seattle), A.KAWAI, M.L.NEUDORFER; Solid-State Electronics v 12 n 3 Mar 1969 p 191-9; Current saturation and oscillations are observed in low resistivity CdSe single crystals without any external illumination over temperature range from room temperature of 77 K; these observations are explained qualitatively by phenomenological theory which is developed on basis of acoustoelectric effect in piezoelectric materials. 28 refs.   10319

Electric Properties. See also Transistors—Field Effect.

Acoustoelectric effects and energy losses by hot electrons—4; A.ROSE (RCA Labs, Princeton, NJ); RCA Rev v 30 n 3 Sept 1969 p 435-74; Semiclassical expressions for spontaneous emission of phonons by energetic electrons are used to derive temperature and field dependence of mobility and mean energy of carriers vs applied field; simple formalism is developed and applied to obtain energy distribution of electrons around their mean energy; comparison of theory and experiment on hot electrons shows that good agreement is obtained for energies near band edge. 22 refs.   09939

Time response of high-field electron distribution function in GaAs; H.D.REES; IBM J Research & Development v 13 n 5 Sept 1969 p 537-42; Techniques used in iterative calculation of steady-state and time-dependent high-field electron distribution function for GaAs, its small-signal frequency response and its behavior in large sinusoidal electric fields; calculations concentrate on frequency range from d-c to about 100 GHz; computed results for response speed, threshold field for negative conductivity, negative mobility, oscillator efficiency, and free-electron dielectric constant. 12277

Space-charge-limited current instabilities in n^+-pi-n^+ silicon diodes; A.K.HAGENLOCHER (General Telephone & Electronics Labs Inc, Bayside, NY), W.T.CHEN; IBM J Research & Development v 13 n 5 Sept 1969 p 533-6; Measurements of instability in nickel-doped 25,000 ohm-cm silicon when strength of pulsed electric field is varied above and below threshold; at threshold nickel centers become ionized, and positive space charge is created in center of sample; this nonequilibrium distribution, which persists for period of 200 to 300 mu sec, has properties similar to those of gaseous plasma. 20 refs. 12278

Films. See also Films—Metallic.

Interface-related electrical properties of cadmium selenide films; R.G.WAGNER (McDonnell Douglas Corp, St. Louis, Mo), G.C. BREITWEISER; Solid-State Electronics v 12 n 4 Apr 1969 p 229-38; Method for measuring depletion depths in thin film semiconductors, described by R.R.Haering, and J.F.O'Hanlon is used to study simultaneous variation of conductivity and depletion depth in thermally processed cadmium selenide films; effects of oxygen adsorption on unprotected semiconductor films and those covered by silicon oxide layer are compared; response of depletion to surface sorption is discussed with regard to experimental data presented. 22 refs. 10377

Epitaxial $BaTiO_3$ on Pt substrate; O.TADA (Tokushima Univ, Japan), Y.SHINTANI, Y.YOSHIDA; J Applied Physics v 40 n 2 Feb 1969 p 498-501; Epitaxial layers of $BaTiO_3$ grow on Pt substrates when—Pt substrate has preferred orientation (111); $BaTiO_3$ powder is sintered on this substrate at 1400 C for 24 hr in air; epitaxial $BaTiO_3$ is single crystalline with orientation parallel to (111) orientation of Pt but with deformation in crystal structure spacing (111) is larger by about 1% (0.023 A) than that of original $BaTiO_3$ which is tetragonal phase at room temperature; remanent polarization and dielectric constant of this specimen are about 1/4 and 1/10 values of those of polycrystalline $BaTiO_3$, respectively. 11507

Structure and electrophysical properties of epitaxial tin chalcogenide thin films; A.G.MIKALAICHUK, D.M.FREIK; Fizika Tverdogo Tela v 11 n 9 Sept 1969 p 2520-5; Structure, optical, and electrical properties of epitaxial tin chalcogenide thin films obtained by vacuum molecular depositions on (100) faces of alkali-halogen crystals; optical conditions of deposition are determined; on basis of optical absorption spectra energy gaps of SnS, SnSe, and SnTe were determined; sign of surface charge was obtained using field effect. 21 refs. In Russian. 12222

LEED study of (0001) face of $YMnO_3$ in order to epitaxy CdSe on this surface; D.ABERDAM (Domaine Univ, Saint-Martin de'Heres, France), G.BOUCHET, P.DUCROS; Surface Science v 14 n 1 Mar 1969 p 121-40; Using technique of low-energy electron diffraction structure of (001) face of yttrium manganite has been studied with view to subsequent epitaxying of sxviconducting cadmium selenide upon this ferroelectric crystal; heat treatments necessary for obtaining ordered surface of $YMnO_3$ have been determined, and possible explanation for appearance of 2x2 surface structure is given; as result of this study epitaxy of CdSe on $YMnO_3$ has been achieved

Appendix C

Polymer Science and Technology —Guides to the Journal and Report Literature

— Selected Abstracts from a
Computer-Based Express Service

6. CELLULOSE AND OTHER CARBOHYDRATES

This section contains digests on the chemical constitution of wood and physical and chemical properties of its constituents. The chemistry and engineering of the paper (and its analogs) industry and viscose for unspecified use are included. Uses for cellulose and its derivatives (except fibers) are covered.

4857   BASIC PRINCIPLES OF GEL FILTRATION. Reiland, John. Amer. Lab. 1969(Oct.), 29-32, 34(Eng).

The dry particle diam., water regain, the bed vol., and mol. wt. fractionation ranges of SEPHADEX, prepd. by crosslinking DEXTRAN with EPICHLOROHYDRIN [106898], are discussed.

The applications and derivs. of Sephadex are reported.

4858   POLYSACCHARIDES. Whistler, Roy L. (Purdue Univ., Lafayette, Indiana). Encycl. Polym. Sci. Technol. 11, 396-424(1969)(Eng).

The chem. structures, sources, and uses of POLYSACCHARIDES such as HEMICELLULOSES, XYLANS, MANNANS, SEAWEED POLYSACCHARIDES, EXUDATE GUMS, and PECTIN are reviewed with 9 refs.

4859   ADSORPTION OF STARCH AND OLEATE AND INTERACTION BETWEEN THEM ON CALCITE IN AQUEOUS SOLUTIONS. Somasundaran, Ponisseril (Int. Min. and Chem. Corp., Libertyville, Ill.). J. Colloid Interface Sci. 31(4), 557-65(1969)(Eng).

The adsorption of STARCH and SODIUM OLEATE [143191] on CALCITE in aq. soln. was studied.

Streaming potential and calcium soly. measurements were made to det. the mechanism of oleate and starch adsorption at the solid-soln. interface.

Source. Post•J, **6**, No. 9, March 16, 1970, pp. 115-116. A publication of the Chemical Abstracts Service, published by the American Chemical Society. Reprinted with permission of the Chemical Abstracts Service of the American Chemical Society.

The results showed the presence of mutual enhancement of adsorption, possibly due to the formation of a helical starch-oleate clathrate, and the interaction of oleate with calcium bound to starch.

4860 DISTRIBUTION OF METHOXYL GROUPS IN THE METHYLATION OF STARCH ALKOXIDES OBTAINED BY METALLATION WITH ALKALI METAL NAPHTHALENE. Ezra, Gabriel; Zilkha, Albert (Hebrew Univ., Jerusalem, Israel). J. Macromol. Sci., Chem. 3(8), 1589-600(1969)(Eng).

Methoxide group distribution during methylation of STARCH was studied by metalating the starch in dimethyl sulfoxide [67685] with alkali metal naphthalenes, treating with methyl iodide [74884], hydrolyzing, and analyzing for glucose and O-methyl glucose derivs.

Formation of 2,3,6-tri-O-methyl-D-glucose [4234440] at low alkoxide concns. and the presence of unreacted glucose at high alkoxide concns. demonstrated occurrence of random metallation.

At low alkoxide concns., preferential metallation occurred at the 2-carbon and 6-carbon hydroxyl groups and increases in the 3-carbon metallation occurred with increasing alkoxide concn.

Methoxy group distribution showed the ease of metallation to be in the order: lithium > sodium > potassium, but the order of methylation efficiency to be potassium > sodium > lithium.

4861 KINETIC STUDY ON THE ALKALINE DEGRADATION OF AMYLOSE. Lai, Yuan-Zong; Sarkanen, K. V. (Univ. of Washington, Seattle, Wash.). J. Polym. Sci., Part C 1969(28), 15-26(Eng).

The degradation of AMYLOSE by 5% SODIUM HYDROXIDE at 100% was studied colorimetrically, by titrn. of acids formed from peeled-off units, or by pptn. with acetone [67641].

The enhancing effect of calcium ions and strontium ions on the termination rate of the peeling-off reaction was rationalized in terms of the promotional effect of these cations in stabilizing the diionic intermediate.

4862 COPOLYMERIZATION OF LEVOGLUCOSAN DERIVATIVES AND UNSATURATED COMPOUNDS. Pernikis, R.; Surna, J.; Orbidane, A. (Inst. Khim. Drev., Riga, USSR). Latv. PSR Zinat. Akad. Vestis, Kim. Ser. 1969(4), 442-5(Russ).

The copolymn. of TRIMETHYLLEVOGLUCOSAN [2951862] and LEVOGLUCOSAN [498077] with STYRENE [100425] and METHYL METHACRYLATE [80626] was investigated.

The copolymn. of trimethyllevoglucosan with styrene was carried out in methylene chloride [75092] at room temp. in the presence of boron trifluoride etherate [109637].

A fraction (A) pptd. from soln. and was repptd. from chloroform [67663] soln. with acetone [67641].

The residual soln. was concd. and a fraction (B) pptd. with diethyl ether [60297].

Increasing the amts. of starting styrene decreased the yield of fraction A, while the yield of B increased from 1.6% to 26.2%.

The viscosity of fraction A for all monomer ratios was const. at 0.090-0.098.

The amts. of unchanged styrene in all cases was 23-24%.

Only small amts. of copolymer were obtained by copolymn. of trimethyllevoglucosan and methyl methacrylate in the presence of boron trifluoride etherate, aluminum chloride [7446700], or tin tetrachloride [7646788].

Attempts to copolymerize levoglucosan, styrene, and methyl methacrylate in dioxane [123911] at 90° in the presence of the above catalysts gave only mixts. of homopolymers.

4863 SWELLING OF CELLULOSE. I. INTRACRYSTALLINE SWELLING AND DECRYSTALLIZATION IN ZINC CHLORIDE-TREATED COTTON CELLULOSE. Betrabet, S. M.; Daruwalla, E. H.; Lokhande, H. T.; Padhye, M. R. (Cotton Technol. Res. Lab., Bombay, India). Cellul. Chem. Technol. 3(4), 309-23(1969)(Eng).

An evaluation was made of the swelling and decrysth. action of aq. ZINC CHLORIDE solns. of different concns. at 20° and 35°.

260

The characteristics of COTTON CELLULOSE, such as d., birefringence, ir crystallinity index, moisture regain, flat bundle strength, extent of swelling, accessibility to acid hydrolysis and formylation, and lateral-order distribution of fiber substance, were altered depending on the concn. of the zinc chloride solns. used.

The intracryst. swelling and decrystn. of cotton cellulose under different conditions of treatment with zinc chloride solns. reflected the nature and extent of formation of sp. hydrates in the aq. soln., rather than to formation of different types of complex between zinc chloride and cellulose.

4864 GRAFTING ON CELLULOSE. Atlas, Sheldon M.; Mark, Herman F. (Bronx Community Coll., Bronx, N. Y.). Cellul. Chem. Technol. 3(4), 325-45(1969)(Eng).

Recent work on the grafting of monomers onto CELLULOSE was reviewed with emphasis on radiation grafting, ozonization grafting, diazotation grafting, redox grafting systems and chain transfer.

45 refs.

4865 OXIDATION OF ALLYLCELLULOSE. Piklerova, Anna; Pasteka, Mikulas; Pikler, Alexander (Slowakischen Akad. Wiss., Bratislava, Czech.). Faserforsch. Textiltech. 20(11), 516-20(1969)(Ger).

The oxidn. rate of ALLYL CELLULOSE at 80-100° in air and in pure oxygen increased with increasing degree of substitution and temp.

The rate of pure oxygen was approx. 2 times that in air, but the activation energy was the same in both cases.

The induction period length decreased with increasing oxygen partial pressure.

Oxidn. caused the disappearance of carbon-carbon double bonds and the appearance of peroxide groups at a rate which passed through a max. and then declined because of increasing peroxide decompn. rate.

Ir spectra provided evidence for the chem. changes during oxidn.

261

Appendix D

Biological Abstracts

— Selected Pages of Abstracts

4057. STEWART, J. (Dep. Genet., Univ., Cambridge, Engl., UK.) Biometrical genetics with one or two loci: 1. The choice of a specific genetic model. HEREDITY 24(2): 211-224. Illus. 1969. --Methods are described which provide a practical and reasonably economical means of testing whether segregation at 1 or 2 loci will account adequately for continuous quantitative variation in any particular back-cross. These methods are based on the comparison of parental, F_1 and back-cross distributions. The validity of these methods is tested empirically by application to some simulated data. The minimum requirements for the application of these methods in summary as follows: there should be 2 inbred strains; the characters to be measured should be continuous variables with a heritability of at least 0·3 in one of the back-crosses; and it should be possible to measure a total of 100 individuals from 3 successive generations. These conditions may be satisfied by a wide variety of organisms used for biological research, and are not restricted to those usually regarded as genetically "well worked". --B. H.

4058. STEWART, J. (Dep. Genet., Univ., Cambridge, Engl., UK.) Biometrical genetics with one or two loci: II. The estimation of linkage. HEREDITY 24(2): 225-238. Illus. 1969. --Methods are presented for obtaining a probabilistic classification of back-cross individuals into genotypic classes, for the case where segregation at 1 or 2 loci can account adequately for the continuous variation in a quantitative character observed in the back-cross. The estimation of linkage relationships between different loci, when such probabilistic classification is available for each locus, is described. The distinction between linkage and pleiotropy is discussed. Possible uses of these methods are described. --B. H.

4059. MERRELL, DAVID J. (Univ. Minn., Dep. Zool., Minneapolis, Minn., USA.) Limits on heterozygous advantage as an explanation of polymorphism. J HERED 60(4): 180-182. Illus. 1969. --A simple table is presented for use in testing the adequacy of the theory of heterozygous advantage as the mechanism of polymorphism. --B. H.

PLANT

See also: Agronomy
Horticulture
Phytopathology - Parasitism and Resistance

† 40060. JONES, GARY E., and ROBERT K. MORTIMER. (Div. Med. Phys., Univ. Calif., Berkeley, Calif., USA.) L-asparaginase-deficient mutants of yeast. SCIENCE (WASHINGTON) 167(3915): 181-182. 1970. --Yeast L-asparaginase is a multimeric enzyme for

which only a single structural gene has been found. Fourteen mutants deficient in L-asparaginase were isolated, and located at one site on the genetic map of Saccharomyces cerevisiae. The L-asparaginase gene (asp1) is located about 18 centimorgans from a gene governing tryptophan synthesis (trp4) on fragment 2 of the map.

† 40061. KHOSHOO, T. N., R. C. MEHRA, and K. BOSE. (Nat. Bot. Gard., Lucknow, Utter Pradesh, India.) Hybridity, polyploidy and change in breeding system in a Ruellia hybrid. THEOR APPL GENET 39(3): 133-140. Illus. 1969. --Ruellia tweediana and R. tuberosa are large flowered chasmogamous diploids (n = 17) with normal meiosis and fertility. F1 hybrids, successful in only one direction (R. tweediana x R. tuberosa), are vegetatively vigorous and possess 17 often heteromorphic bivalents with high degree of segregational irregularities. It is exclusively cleistogamous and completely pollen and seen sterile. Like F1, the artificial amphidiploid (n = 34) is also cleistogamous but shows preferential chromosome pairing with complete restoration of fertility. The parental chromosomes are sufficiently differentiated and cleistogamy is either genic or due to gene-cytoplasm interaction but sterility is entirely chromosomal. All floral parts excepting calyx are highly deformed. Such a deformity is associated with sterility in the F1 but with fertility in the amphidiploid. This is perhaps the 1st case of origin by hybridization of a true breeding and fully fertile cleistogamous taxon from 2 chasmogamous species. It also shows the extent and nature of change in breeding system brought about by hybridization and/or polyploidy. The chromosome numbers in the 6 out of 16, obligate cleistogamous taxa show that they are high polyploids. Perhaps their origin has been in the same manner as in the present case.

† 40062. ROD, J. (Inst. Ackerbau, Abt. Feldversuch. und Biometrie, Brno, Czech.) Beitrag zur Schatzung der allgemeinen Kombination-selgrung von Luzerneklonen. [A contribution to an evaluation of general combining ability of alfalfa clones.] THEOR APPL GENET 39(3): 127-132. Illus. 1969. --Recombinational performance of different clones of alfalfa was compared in 2 types of seed: produced by systematic mating of the clones and produced by the polycross procedure. No basic differences between the 2 types of progeny were found for the characteristics investigated. Thus neither greater objectivity nor greater test efficiency of the polycross technique was demonstrated. Factors determining objective evaluation of general combining ability of alfalfa by means of the polycross test technique are discussed.

† 40963. CAROLINE, DINA F. (Genet. Found., Univ. Tex., Austin, Tex., USA.) Pyrimidine synthesis in Neurospora crassa gene-enzyme relationships. J BACTERIOL 100(3): 1371-1377. 1969. --A new series of pyrimidine-requiring mutants of Neurospora was isolated and all enzymes involved in pyrimidine biosynthesis are represented by at least one mutant. Among these mutants is included a single isolate for a new locus, pyr-6. This mutant is deficient in dihydro-orotase (DHOase) and represents the only enzymatic step in orotate synthesis for which no mutant previously had been found. This mutant, which mapped genetically on the right arm of linkage group V, is unlinked to any of the other pyrimidine mutants. The DHOase-deficient mutant is also characterized by an unexpected growth behavior. The pyr-1 locus is specifically associated with a lack of dihydroorotate dehydrogenase (DHOdehase). Mutants isolated in this series for other pyrimidine loci have been related to previously isolated mutants by allelism, recombination, and accumulation studies.

† 40964. KENDALL, W. A., and N. L. TAYLOR. (Dep. Agron., Univ. Ky., Lexington, Ky., USA.) Effect of temperature on pseudo-self-compatibility in Trifolium pratense L. THEOR APPL GENET 39(3): 123-126. 1969. --A relatively high temperature treatment, applied during anthesis, was shown to enhance self-seed production through pseudo-self-compatibility in normally self-incompatible red clover (Trifolium pratense L.). The self-seeds were produced in cultures of excised stems held in 2.5% sucrose. The stems were excised when petal color was beginning to appear in the buds. During anthesis the cultures were incubated with the flower heads at 40° and the stems at 25°C. When most of the florets per head had opened the cultures were transferred to 20°C and held at that temperature during the period of pollen growth through the styles and also during seed development. The addition of calcium nitrate and boric acid to the culture medium did not enhance anthesis, seed weight, or the number of seeds produced. Plant genotype and the environment provided before anthesis were the primary factors influencing the number of self-seed produced. Although not all attempts to produce self-seed have been successful, with repeated trials all clones we tested produced some seed.

† 40965. NAYAR, N. M. (Cent. Potato Res. Inst., Simla, Himachal-Pradesh, India.) Considerations on overcoming intrasomatic selection during mutation breeding of vegetatively propagated plants. THEOR APPL GENET 39(3): 99-103. 1969. --When seeds and vegetative tissues are irradiated, a competition occurs in meristematic tissues between lethally and sublethally affected cells on one hand and unaffected cells on the other hand to the advantage of the latter. This phenomenon was first recognized by Freisleben and Lein (1943) and termed intrasomatic selection by Kaplan (1951) and diplontic selection by Gaul (1957). The work done on the nature of this phenomenon in seed-propagated plants is summarized along with suggested methods to reduce its intensity. The generally unencouraging results obtained so far in vegetatively propagated plants have been attributed to the more intensive operation of this phenomenon in this group of plants, consequent on the more complex nature of apical meristems and propagating materials. The work of Zwintzscher (1959) in apples and Bauer (1957) in black currants have shown that it is possible to overcome its effects with suitable handling of irradiated materials. The radiogenetic studies done in the potato have been reviewed in this context. Low intensity irradiation, dose fractionation and irradiation of physiologically dormant tubers gave higher survival value. The method of isolating sprouts from irradiated tubers and growing them, and scoring for mutations in sprouting progenies was found to increase mutation frequency considerably. Other methods proposed for minimizing the effects of intrasomatic selection in vegetatively propagated plants are chronic irradiation, neutron irradiation, chemical mutagenesis and lastly, mutagenesis of isolated single cells and growing them into whole plants.

† 40966. BLAICH, ROLF, and KARL ESSER. (Inst. Allgem. Bot., Ruhr-Univ., Bochum, West Ger.) Der cis-trans-Positionseffekt in heterokaryon morphologischer Mutanten von Podospora anserina. [The cis-trans position effect in heterokaryons of Podospora anserina morphological mutants.] ARCH MIKROBIOL 68(3): 201-209. Illus. 1969. --The cis-trans position effect of the 2 non-linked morphological mutants z and oct-1 in heterokaryons of P. anserina is caused by quantitative differences of the 2 types of nuclei. In the cis configuration, there is a growth diminution of hyphae containing predominantly nuclei of the double mutant type which leads to a selection of hyphae containing mainly the wildtype nuclei. This is due to the formation of microsectors. After 6 cm of mycelial growth the double mutant nuclei are no longer distinguishable; the mycelium adopts the wild phenotype. In the trans configuration, the microsectors of both types of mutant nuclei exhibit the same growth rate; the separation of nuclei is prevented as a result of the formation of hyphal anastomoses. The mycelium shows then an intermediate phenotype with both mutant characteristics.

264

according to whether the 2 enzymes are extracted from euryhaline crustacean muscles (Carcinus maenas, Eriocheir sinensis, Astacus fluviatilis) or from stenohaline crustacean muscles (Homarus vulgaris, Maia squinado). The results were discussed in the scheme of the amino-acid metabolism. The difference between an euryhaline and a stenohaline crustacean consists of a control by intracellular liquid salt concentration of the activity of certain enzymatic systems playing a role in the amino-acid catabolism.--S. A.

46460. KRISHNAMOORTHY, R. V. (Dep. Zool., Sri Venkateswara Univ., Tirupati, Andhra Pradesh, India.), and V. VENKATA REDDY. Hepatopancreatic amylase activity as a function of warm-adaptation in a fresh-water field crab. EXPERIENTIA 24(10): 1019-1020. Illus. 1968. [Ger. sum.]

46461. DE SOUZA, TEREZINHA TAVARES, and MARIA DA CONCEICAO CALAND. (Fac. Farm. Bioquim. Univ. Fed., Ceara, Brazil.) Estudo preliminar sobre a bacteriologia do caranguejo Ucides cordatus (Linnaeus). [Preliminary study on the bacteriology of the crab Ucides cordatus Linnaeus.] ARQ ESTAC BIOL MAR UNIV FED CEARA 8(1): 107-108. 1968[recd. 1969]. [Engl. sum.]-- This paper is a preliminary contribution toward a better understanding of the bacteriology of U. cordatus from the coast of Ceara State, Brazil. The species Bacillus pumilus Gottheil and Achrombacter delicatulum (Jordan) were found living in the intestine of the crab. Another species was not identified.--D. M. G.

46462. ALVES, MARIA IVONE MOTA, and ROBERTO CLAUDIO F. BEZERRA. (Estacao Biol. Marinha, Univ. Fed. Ceara, Ceara, Brazil.) Sobre o numero de ovos da lagosta Panulirus argus (Latr.). [On the number of eggs of the spiny lobster Panulirus argus Latr.] ARQ ESTAC BIOL MAR UNIV FED CEARA 8(1): 33-35. Illus. 1968[recd. 1969]. [Engl. sum.]--A study was made of the relationship of total length to number of eggs carried by the spiny lobster P. argus, which lives along the coast of the State of Ceará (Brazil). The following conclusions were drawn: The number of eggs carried by spawned females depends upon their own length. The equation that represents the total length to number of eggs relationship is

$$E = 4.8 \ L \ 3.53,$$

where, E = number of eggs, L = total length, a = constant, and b = exponent. The parameters a and b were calculated through a regression fitted by least squares of the logarithmic transformation:

$$\log E = \log a + b \log L.$$

--D. M. G.

46463. RANGNEKER, P. V., and M. N. MADHYASTHA. (Anim. Physiol. Div., Inst. Sci., Bombay, Maharashtra, India.) Effect of eyestalk removal on the rate of oxygen consumption in the crab, Scylla serrata (Forskal). J ANIM MORPHOL PHYSIOL 16(1): 84-88. 1969.--The effect of eyestalk ablation on O_2 consumption in the crab S. serrata is determined. Twenty-four hr after eyestalk removal there is a rise in O_2 consumption in destalked animals. Injection of eyestalk extract brings about almost complete restoration of the rate of O_2 consumption in the animals, but sinus gland implantation in destalked crabs has practically no effect. From these observations, it is suggested that the rise in the rate of O_2 consumption observed in the destalked crabs is related to the deprivation of the X-organs in the eyestalks which are the sites for the synthesis of the factor responsible for regulating the rate of O_2 consumption.--D. M. G.

46464. MAKAROV, P. O., and Z. E. GUTLAND. (A. A. Zhdanov Leningrad State Univ., Leningrad, USSR.) Ritmogenez v odinochnom nervnom volokne chernomorskogo kraba Eriphia spinifrons. [Rhythmogenesis in a single nerve fiber of the Black Sea crab (Eriphia spinifrons).] BIOFIZIKA 14(5): 858-865. Illus. 1969. [Engl. sum.]--The following change of the repeated discharge obtained while stimulating a single pulpless nerve fiber with short right-angled impulses of the electric current was observed for 5-7 hr: the number of peaks decreased, the intervals between them increased, and there was a tendency toward decrease of the discharge duration. The stimulus intensity being 16-24 rheobases, the shortest interval was between the 2nd and 3rd impulse; the increase of this impulse during the experiment was less than that of the others. When the stimulus intensity increased from 7 to 50 rheobases, various changes of the intervals between the peaks were observed: when the stimulus is increased up to 20 rheobases the intervals between the 1st and 2nd peaks decreased on the exponent to 4 msec; further strengthening of the stimulus up to 35 rheobases caused a sharp increase of the interval up to 16 msec, and smooth decrease to 10 msec; further increase of the stimulus brought about the leap of the interval up to 19 msec and its almost linear decrease up to 13 msec; the interval between the 2nd and 3rd peaks decreased exponentially with the strengthening of the stimulus up to 32 rheobases, then was not changed and remained close to 6 msec; the 4th peak appeared only at high intensities of the stimulus beginning with 35 rheobases. When the stimulus was strengthened the interval between the 3rd and the 4th peak decreased to 8 msec, after which it increased.

46465. ABOLMASOVA, G. I. (Inst. Biol. S. Seas, Acad. Sci. Ukr. SSR, Sevastopol, USSR.) Intensivnost' dykhaniya kraba-vodolyuba Xantho hydrophilus (Herbst). [Respiration rate of the crab Xantho hydrophilus (Herbst).] GIDROBIOL ZH 5(4): 114-117. Illus. 1969.-- The relationship between the respiration rate and the body weight of the Black Sea crab may be described by the following coefficient, $a = 0.087$ and $k = 0.77$, where a = the coefficient describing overall metabolism in a crab with unit body weight, and k = the constant relating change in the rate of metabolism with weight. The average consumption of O_2 by crabs weighing 0.9 to 16.5 g is 0.084-0.691 ml/crab/hr. The crabs were calculated to expend 0.41-3.36 cal/crab/hr.--R. W.

MYRIAPODA

†46466. WEATHERSTON, J., and J. E. PERCY. (Insect Pathol. Res. Inst., Dep. Fish. and Forest., Sault Ste. Marie, Ont., Can.) Studies of physiologically active arthropod secretions: III. Chemical, morphological and histological studies of the defence mechanism of Uroblaniulus canadensis (Say) [Diplopoda Julida]. CAN J ZOOL 57(6): 1389-1394. Illus. 1969.--The defensive secretion of a millipede, U. canadensis, is shown to consist of a mixture of benzoquinones and aliphatic compounds. Two of the benzoquinones are identified as p-benzoquinone and 2,3-dimethoxybenzoquinone. The gross structure of the gland producing the secretion is described and histological studies reveal the presence of secretory units. Each secretory unit consists of a pair of specialized cells and has a tubule to transport the secretion manufactured in these cells to a large cuticular-lined reservoir.

†46467. BRIMHUBER, BARBARA S. (Zool. Dep., Univ., Cape Town, S. Afr.) The mode of sperm transfer in the scolopendromorph centipede: Cormocephalus anceps anceps Porat. ZOOL J LINNEAN SOC 48(3): 409-420. Illus. 1969.--The laboratory conditions under which the centipedes were kept and their general behavior are described. No indication of sexual recognition between adult centipedes was found during studies made of encounters between members of the same or opposite sex. All displayed the same avoiding reaction on the first encounter and later adopted the defence posture by which fatal attacks were prevented. Where attacks were not successfully countered, biting may or may not be accompanied by injection of poison into the wound. These centipedes are not immune to the poison of their own species, therefore, fatalities and cannibalism can result from aggressive encounters. The mating behavior of Cormocephalus anceps was witnessed once in early autumn in South Africa. It was preceded by the typical avoiding reaction and defence posture assumed by either sex towards the other, the male being the more dominant. After adopting this posture, a situation which may last a variable length of time, the subsequent stages of mating behavior were initiated by the male. The courting behavior could be followed in 5 sequences which are described in some detail. The problems of sperm transfer faced by the Arthropoda evolving towards a wholly terrestrial existence and the many modes adopted by this group are considered together with the recent knowledge of the modes of sperm transfer employed by the four orders of Chilopoda.

INSECTA

See also: Economic Entomology
General and Systematic Zoology - Invertebrata - Insecta
Phytopathology - Diseases Caused by Animal Parasites, Disease Control
Public Health - Disease Vectors, Animate, Disinfection and Vector Control

General

†46468. BENNETT, LEON. (Sch. Eng. and Sci., N. Y. Univ., New York, N. Y., USA.) Insect flight: Lift and rate of change of incidence. SCIENCE (WASHINGTON) 167(3915): 177-179. 1970.--Large changes in lift output result when a stimulated insect wing, undergoing a downstroke, is subjected to a dynamic change of incidence. Given a large positive rate of change of incidence, transient lift values several times those realized in steady-state operation at the same angle of incidence are obtained. Thus, a means exists by which insects achieve several times the lift expected by conventional quasi-steady considerations.

†46469. VARMA, M. G. R., and MARY PUDNEY. (London Sch. Hyg. and Trop. Med., London, Engl., UK.) The growth and serial passage of cell lines from Aedes aegypti (L.) larvae in different media. J MED ENTOMOL 6(4): 432-439. Illus. 1969.--The establishment and growth of cell lines from larval A. aegypti in different media, without insect hemolymph, are described. The method of subculturing the cells, using a 0.05% solution of pronase instead of trypsin solution to detach the cells from the glass surface, was simple, quick and easily reproducible. There were differences in the architecture of the cell layer and of the predominant cell type in the cell lines. One of the cell lines which was studied in greater detail had an estimated population-doubling time of 29 hr during the active growth phase and showed a 25-fold increase in cell number during 7 days of growth. The majority of cells in this line retained the normal 2n (= 6) complement of

Appendix E

Chemical Abstracts

— Selected Pages of Abstracts

30445c **Egg white, free of glucose, containing an inhibitor preventing discoloration.** Consolidated Foods Corp. **Brit.** **1,170,318** (Cl. A 23b), 12 Nov 1969, US Appl. 20 Feb 1967; 5 pp. In order to prevent discoloration of liquid egg white, frequently occurring during the prolonged heat treatment used to kill *Salmonella*, egg white is processed to remove its glucose content to a level that upon drying the dried product contains ≦0.2% by wt. glucose. Then an Al salt (acetate, chloride, phosphate, sulfate of Al, or K Al tartrate, Na Al chloride, NH₄ Al sulfate or Na Al sulfate) is added in amts. of 0.35–0.75% by wt. (expressed as Al³⁺), referred to egg white. The point of addn. of Al³⁺ is not crit. but it is essential that the egg white be liq. on addn. Istvan Finaly

30446d **Two-component extract of garlic.** Lazarev, I. Z.; Ivanova, O. I. (Krasnodar Scientific-Research Institute of Food Industry) **U.S.S.R. 244,879** (Cl. A 23bl), 28 May 1969, Appl. 27 Feb 1968; From *Otkrytiya, Izobret., Prom. Obrazlsy, Tovarnye Znaki* 1969, 46(18), 165–6. Garlic is cooled to inactivate enzymes, ground, and the enzymes are removed from the ground mass by aq. extn., followed by the prepn. of an enzymic ext. (1st component) which is sepd. from the remaining mass. The latter component is extd. with liq. CO₂ to produce an enzyme-free ext. (2nd component). The 2 exts. are mixed together before they are added to foods. MSCL

30447e **Artificial fish foods.** Zitserman, M. Ya.; Polunina, N. I.; Belen'kii, B. M.; Sivertsov, A. P.; Polstyanko, T. P. (All-Union Scientific-Research Institute of Grain and Grain Products and All-Union Scientific-Research Institute of Reservoir Fish Husbandry) **U.S.S.R. 244,795** (Cl. A 01k, A 23k), 28 May 1969, Appl. 15 Mar 1968; From *Otkrytiya, Izobret., Prom. Obrazisy, Tovarnye Znaki* 1969, 46(18), 143. Artificial fish foods are obtained by prepg. and granulating a food mixt. To influence the water resistance of the granules and preserve the nutrients of the food, the food mixt. is treated with a surfactant, such as 5% aq. lignosulfonate, during granulation. MSCL

30448f **Baking additive.** Cooke, Alfred; Johnnson, Harold (Delmar Chemicals Ltd.) **S. African 68 02,093** 28 Mar 1969, US Appl. 03 Apr 1967; 22 pp. The addn. of small amts. of a pentosan-degrading enzyme (I) prepn. into flour results in a lasting softening and whitening effect on the baked bread. The max. amt. of I to be added is inversely proportional to the fermentation time. Thus, 5 loaves were prepd. using the std. sponge-

dough procedure and contg. 0, 0.0025%, 0.005%, 0.01%, and 0.015% by wt. of Rhoyzme HP-150 (1800 units of pentosanase activity/g). The baked loaves were increasingly softer with degree of concn. of the enzyme. Ellen T. Dec

30449g **Oil and feed cake from tanned vegetable sources.** Institut National de la Recherche Agronomique and Produits Chimiques et Celluloses Rey **Fr. 1,523,268** (Cl. C 11b), 03 May 1968, Appl. 10 Mar 1967; 4 pp. Addn. of com. tannant to a seed prepn. after husking, before rolling or pressing, improves oil production through better percolation and demucilaging, and prevents bacterial decompn. of protein in the cake. Any tannin mixt. may be used, but it should be pure enough to yield sufficient tannin without adulterating the feed cake. For best results, add to peanuts 15, sunflower seed 15, rapeseed 12, soybean 15, and linseed 6% tannin, at 70–80° and 10–15% moisture. E. L. London

30450a **Artificially sweetened fruit-flavored product.** Block, Harry W.; Adams, Joan M.; Finucane, Thomas P. (General Foods Corp.) **U.S. 3,476,571** (Cl. 99–130; A 23j), 04 Nov 1969, Appl. 22 Aug 1963-19 May 1966; 3 pp. A sweetener is prepd. from cyclamate, saccharin, mannitol and a bland hydrophilic colloid; 1–10 parts by wt. cyclamate are used per part saccharin and 1–30 parts mannitol and 0.8–1.2 parts colloid are present per part cyclamate and saccharin combined. This sweetener may be used with fruit flavoring and an org. food acid to produce a dry, free-flowing, powdery edible compn. of desirable tartness and sweetness, free from undesirable aftertaste. Thus, a sweetener was prepd. by dry-blending mannitol 10.0, gum arabic 1.1, Na cyclamate 1.0 and Na saccharin 0.3 parts by wt. Then it was used to prep. a low-cal. gelatin dessert as follows: sweetener 39.6, gelatin 51.2, citric acid 6.0, trisodium citrate 1.4, lemon flavor 1.0 and color 0.8 parts. When 6–7 g of the dry mix compn. was added to 237 ml 180° water to form a clear soln. and then allowed to gel, the gelatin had a sweet, tart taste with no residual bitter aftertaste as compared to a control prepd. using only the

Source, Chemical Abstracts, **72,** No. 7, February 16, 1970. Reprinted with permission of the Chemical Abstracts Service of the American Chemical Society.

saccharin and cyclamate which had a bitter, medicinal, metallic aftertaste. Ellen T. Dec

30451b **Choline composition.** Jones, Lawrence R. (Commercial Solvents Corp.) **U.S. 3,475,177** (Cl. 99–2; A 23k), 28 Oct 1969, Appl. 09 Feb 1966; 3 pp. A compn. resistant to caking, contg. a choline salt with a solid fat, is prepd. by dissolving the fat in a liq. fat solvent, dispersing the choline salt in the fat solvent (0.33–19 parts by wt. per part of fat). The combination is recovered from the fat solvent in particulate form. Thus, to 320 g anhyd. iso-PrOH at 60° was added 15 g flaked tallow and 35 g anhyd. choline chloride crystals. When the tallow and crystals were dissolved, the soln. was cooled in an ice bath. The ppt. was filtered and dried in a vacuum oven at 40°. The combination sought was 70% choline chloride, 30% tallow. By assay the combination obtained was 72% choline chloride and 28% tallow by wt. The choline chloride–tallow combination was used in a choline supplement for use in an animal feed compn. and the tallow was included in calcg. fat content. The supplement contg. the choline chloride–tallow combination showed little tendency to cake during storage. Similarly formulated supplement with choline chloride contg. no tallow caked seriously.
S. P. Marino

30452c **Product in powder form containing chlorine dioxide.** Lovely, Clement F. (International Dioxcide Inc.) **Ger., Offen. 1,911,995** (Cl. C 01b), 06 Nov 1969, US Appl. 16 Apr 1968; 19 pp. The product is mixed with citric acid powder to liberate ClO_2 for conservation of fruits and vegetables in transport and storage. $FeCl_3$ and acids are used to set the gas free as disinfectant or deodorant and for sterilization. An aq. soln. of ClO_2 is stabilized with $Na_2CO_3 + H_2O_2$ (pH 8.5–9) and taken up by talc or synthetic calcium silicate which has a pH 8.3–8.6 forming a dry powder. Acidification releases ClO_2. The acidifier consists either of citric acid or $FeCl_3$ soln. (pH 2–6) and is absorbed by silicic acid powder. The ratio of release of the gas is controlled by the amt. of acidifier added and the pH. The rate of release can be 0.001%–1.0% in reference to the total vol. of the storage container.
Hugh H. Wartell

30453d **Aerosol-dispensed compositions for the preparation of thick shakes and puddings.** Diamond, George B. **Brit. 1,169,169** (Cl. A 23l), 29 Oct 1969, US Appl. 22 Oct 1965; 8 pp. Thick shakes and puddings can be prepd. from a syrup dispersed from an aerosol container and comprising an aq. suspension of a flavoring agent and 3–8% of an alginate of viscosity ~10 cp in a

1% soln. of water at 68°F. This syrup has a viscosity <7000 cp at 40° and when added to 7 parts milk forms a thick shake and in 5 parts milk a soft pudding. Thus, a syrup of 1400 cp (40°) was formed from: cocoa 5.0, invert sugar 45.0, emulsifier 0.04, salt 0.06, vanilla flavor 0.13, carrageenin 0.30, Na alginate 3.0 lb and water to make 100 lb. When 1 part of this syrup was injected through a venturi nozzle into 7 parts milk, a chocolate milk shake of viscosity 600 cp was formed.
Ellen T. Dec

30454e **Monoglyceride emulsifiers for foods.** Swicklik, Leonard J. (Eastman Kodak Co.) **Fr. 1,537,452** (Cl. A 23l), 23 Aug 1968, US Appl. 22 Sep 1966; 5 pp. The emulsifiers contain C_{16-20} fatty acid monoesters of glycerol and 1,2-propanediol. At least 75% of the fatty acids in the esters of both glycols should be the same and there should be <5% unsatd. acids. The ratio is 35–60% glycerol esters and 40–65% propanediol esters. The emulsifier also contains hydroxylated lecithin, an antioxidant, a preservative, and water. Thus, an oil phase is obtained by mixing 5 kg Myvatex type 3–50 and 200 g hydroxylated lecithin (Centrolene S), and heating at 71°. The aq. phase is obtained by mixing 14.76 kg H_2O and 40 g K sorbate and heating at 71°. While stirring, the oil phase is added to the aq. phase and cooled to 51–3° with stirring. The emulsifier is used to prep. dairy, vegetable, and fruit products with the consistency of whipped cream, with addn. at the 2–8% level.
Harry De Moor

30455f **Inhibiting oxidation of fats with 5-hydroxytryptophan.** Ito, Hiroshi (Tanabe Seiyaku Co., Ltd.) **Japan. 69 25,699** (Cl. 19 A 1), 29 Oct 1969, Appl. 17 Feb 1966; 3 pp. Of tryptophan, histidine, 5-hydroxytryptophan (I), and NDGA I is the most effective in inhibiting oxidn. of fats and fatty foods. Thus, 140 mg linoleic acid (II) dissolved in 0.5 ml EtOH is added to 4 ml of $0.1 M$ buffer (KH_2PO_4-NaOH, pH 7.0). When 10^{-6} mole I is added to the II soln., the II soln. shows 97 and 93% residual II (detd. by polarography with lipoxidase), as compared with 97 and 84% with 10^{-4} mole NDGA.
Hideo Kuroe

28253v Recent developments in molecular taxonomy. Erdtman, Holger (Roy. Inst. Technol., Stockholm, Swed.). *Perspect. Phytochem.*, *Proc. Phytochem. Soc. Symp.* **1968** (Pub. 1969), 107–20 (Eng). Edited by Harborne, J. B. Acad. Press: London, Engl. A review with 46 references.
John Christensen

28254w Bibliography on the radiation chemistry of alkanes. Potvin, Eve M.; Ross, Alberta B. (Univ. of Notre Dame, Notre Dame, Indiana). *U.S. At. Energy Comm.* **1969**, COO-38-661, 68 pp. (Eng). Avail. Dep.; CFSTI. From *Nucl. Sci. Abstr.* **1969**, 23(17), 33205. Literature on the radiation chemistry of alkanes is covered by this bibliography. Investigations of aq. solns. of alkanes are not included. The papers are listed by serial nos. that are approx. chronological under the following categories: gas phase, liq. phase, solid phase, heterogeneous systems, radiation synthesis, oxidn., and books and reviews. An author index and compd. index are included.
TCNG

28255x History and geography of alcoholic beverages. Bernabai, Adalberto. *Ann. Sanita Pubb.* **1968**, 29(6), 1555–92 (Ital). Some etymol. explanations, the continental distribution and geographical location, the botany of the productive plants, the lore, history, and prepn. of the alc. beverages from plants and milk are presented. The basic fermentation reactions and the compn. of some kinds of kumiss and kefir are given.
Giuseppe Bini

28256y Recollections of personalities involved in the early history of American biochemistry. Rose, William C. (Univ. of Illinois, Urbana, Ill.). *J. Chem. Educ.* **1969**, 46(11), 759–63 (Eng). Biographies are given of the scientific careers of Samuel Johnson, Russel Chittenden, and Lafayette Mendell with emphasis on their talents and influence which made Yale a leader in the training of biochemists in the U.S. from the inauguration in 1874 of the first course in physiol. chem. until 1915.
Wm. MacL. Pierce

28257z Historical origins of organometallic chemistry. II. Edward Frankland and diethylzinc. Thayer, John S. (Univ. of Cincinnati, Cincinnati, Ohio). *J. Chem. Educ.* **1969**, 46(11), 764–5 (Eng). Edward Frankland (1825–1899), a student of Bunsen, reported the first direct syntheses of organometallic compds. in 1849. He found that Zn, Sn, Sb, As, and P reacted with Et-, Pr-, Bu-, amyl-, and Ph-I. The use of these R_2Zn compds. has been largely replaced by the organo-Mg compds. except in the Simonds-Smith prepn. of cyclopropanes.
Wm. MacL. Pierce

28258a Perkin's original alizarin and mauveine after 100 years and more. Gurr, Edward (Michrome Lab., Edward Gurr Ltd., London, Engl.). *J. Soc. Dyers Colour.* **1969**, 85(10), 473–4 (Eng). Spectral curves are presented for samples of the original alizarin (C.I. 58,000) synthesized by Perkin in 1869 and of manuveine (C.I. 50,245) synthesized in 1856.
Fred H. Steiger

28259b Development of a thermal method for analyzing the solidification and cooling of castings. Bilyk, V. Ya. (USSR). *Tr. Soveshch. Liteinym Svoistvam Splavou, 1st 1966* (Pub. 1968), 16–31 (Russ). Edited by Nekhendzi, Yu. A. Inst. Probl. Lit'ya Akad. Nauk Ukr. SSR: Kiev: USSR. A historical review is given, discussing the foundation and the development of thermal anal. and its application in metallurgy. 45 refs.
Kalojan R. Manolov

28260v One-hundredth anniversary of Mendeleev's discovery of the periodic law and system of chemical elements. Markevich, S. V. (USSR). *Vestsi Akad. Navuk Belarus. SSR, Ser. Khim. Navuk* **1969**, (4), 5–10 (Russ). Mendeleev's discovery of the periodic law is commemorated.
A. J. McHugh

28261w Modern methods for teaching the fundamentals of ionic theory. Langner, Alicja; Langner, Marian; Kieszkowski, Rudolf (Poland). *Chem. Szk.* **1967**, 13(4), 150–9 (Pol). The new syllabus of chemistry in the 2nd grade of the Lyceum recommends teaching the fundamentals of ionic theory on the basis of b.p. and f.p. measurements or by examg. the osmotic pressure of solns. of electrolytes and nonelectrolytes. Better results were obtained by detecting the passage of elec. current between Cu electrodes on diln. of solns. of electrolytes or on dissolution of ionic compds. in polar solvents. The above expts. enabled students to learn the effect of solvents on dissocn. of ionic compds. and helped to eliminate the common mistake in attributing the dissocn. phenomenon to the passage of elec. current.
J. L. Kornacki

28262x Difficulties in learning chemistry. Szymanczyk, Edward (Poland). *Chem. Szk.* **1967**, 13(4), 160–9 (Pol). The great-

est difficulties in learning chemistry were encountered among those primary school pupils who were unable to perform simple expts. in the school lab. and could not visit chem. plants. Screening scientific films and slides, showing various models, tables, and diagrams, discussing popular scientific books, as well as organizing chemistry practicals, should help the teacher in developing pupil interest in chemistry.
J. L. Kornacki

28263y Development of student interest in chemistry. Liebert, Irena (Poland). *Chem. Szk.* 1967, 13(4), 169–72 (Pol). The local school Olympics at the Gostyn Lyceum is described. A set of problems, prepd. for the participants, deals not only with topics of the school chemistry syllabus but also with problems discussed in the press, in popular scientific literature, on the radio, television, etc. The Chemical Olympics develops student interest in chemistry.
J. L. Kornacki

28264z Remarks on the Chemical Olympics [high school chemical competitions]. Mitera, Tadeusz (Poland). *Chem. Szk.* 1967, 13(4), 175–7 (Pol). The standard of the Chemical Olympics improved considerably during the last decade but the no. of competitors decreased. Presumably, primarily those who are helped by their relatives and those who are members of the school chemistry clubs can cope with the Olympics program. The publication of lists of references on which the Chemical Olympics will be based in a given year, as well as sending of marked papers back to the schools, would encourage more pupils to take part in the Olympics and would improve the std. of the preparatory work in the school.
J. L. Kornacki

28265a Projection of colloidal chemistry experiments. Nedzynski, Lucjan (Poland). *Chem. Szk.* 1967, 13(4), 177–84 (Pol). Simple test tube expts. are described which can be screened in a classroom by means of a slightly modified epidiascope. They include prepn. of colloids by pptn., redn., oxidn., and hydrolysis, and coagulation, peptization, and dialysis of colloids.
J. L. Kornacki

28266b Radiation chemistry. Stolarczyk, Lech (Poland). *Chem. Szk.* 1967, 13(5), 200–11 (Pol). The sources of radiation, mechanism of transfer of radiation energy, chem. irradn. results, research methods applied, as well as the role and perspectives of further development and applications of radiation chemistry were described.

28267c Chemistry teaching programs in nonchemical technical schools. Paciorek, Alojzy (Poland). *Chem. Szk.* 1967, 13(5), 211–20 (Pol). The new chemistry teaching program in nonchem. tech. secondary schools is extensively reviewed. The new program is better adapted to the mental capacities of students than the previous one. It covers all topics which are important for liberal education and which form a good basis for prepg. for a profession.
J. L. Kornacki

28268d Repetition in chemistry lectures. Jaskiewicz, Nikodem (Poland). *Chem. Szk.* 1967, 13(5), 220–4 (Pol). Review classes are in most cases combined with marking of the pupils progress but such an approach does not contribute to better understanding of chemistry. Review classes should be fully devoted to discussion of difficult topics, if possible with screening of scientific films and with reference to popular scientific books and radio broadcasts. Introduction of chem. calcns. may also be of great help.
J. L. Kornacki

28269e Simple experiments. Gorzynski, Stanislaw (Poland). *Chem. Szk.* 1967, 13(5), 225–7 (Pol). Simple classroom demonstrations are described. They involve prepn. of Na and K peroxides, reducing properties of Na and K, reaction of Al with air O, and reducing properties of Mg.

28270y Simple laboratory aids. Weglowski, Zbigniew (Poland). *Chem. Szk.* 1967, 13(5), 228–9 (Pol). Empty drug glass containers and their polythene stopcocks can be utilized in a school lab. Pupils can easily adapt them for chem. purposes, thus developing useful manual skills. Tourist gas bottles can be easily fitted with tubes of the Teclu or Mekker burners and serve as lab. gas burners.
J. L. Kornacki

28271z Summer courses at Brodnica for chemistry teachers. Kubicki, Jozef (Poland). *Chem. Szk.* 1967, 13(5), 233–4 (Pol). Summer courses organized at Brodnica for chemistry teachers from primary and secondary schools were described.
J. L. Kornacki

zx49d **Enzyme-substrate complex between valyl ribonucleic acid synthetase and ribonucleic acid specific for valine.** Lagerkvist, Ulf (Univ. Gothenburg, Gothenburg, Swed.). *Struct. Funct. Transfer RNA 5S-RNA, Proc. Meet. Fed. Eur. Biochem. Soc. 4th* 1967 (Pub. 1968), 143-9 (Eng). Edited by Froeholm, L. O. Acad. Press: London, Engl. The isolation with some properties of a stable enzyme-substrate complex involving valyl-RNA synthetase and tRNA specific for valine is described. RCTT

28499e **Inhibition of urease by hydroxamic acids.** Gale, Glen R.; Atkins, Loretta M. (Veterans Admin. Hosp., Charleston, S.C.). *Arch. Int. Pharmacodyn. Ther.* 1969, 180(2), 289-98 (Eng). A series of 61 hydroxamic acids, including sorbylhydroxamic acid, glutarylmonohydroxamic acid, benzoylhydroxamic acid, 4-fluorobenzoylhydroxamic acid, isonicotinylhydroxamic acid, and 2,3-dichlorophenoxyacetylhydroxamic acid, was assessed for inhibitory activity against urease. Twenty-one of these compds. appreciably inhibited the cell-free enzyme from jack bean and from *Proteus morganii* at pH 5.8. Twenty of them traversed the bacterial membrane of *P. mirabilis* at pH 8.0 and inhibited the enzyme intracellularly. Kinetic studies showed that the aliphatic hydroxamic acids were noncompetitive inhibitors, while the aryl derivs. were of a mixed type. Values of K_m/K_i ranged from 100 to 6000. BQJN

28500y **Hormonal and nonhormonal control of glycogen synthesis. Control of transferase phosphatase and transferase I kinase.** Larner, Joseph; Vallar-Pallasi, Carlos; Goldberg, Nelson D.; Bishop, Jonathon S.; Huijing, Fr.; Wenger, J. I.; Sasko, H.; Brown, Norman Edward (Coll. of Med. Sci., Univ. of Minnesota, Minneapolis, Minn.). *Contr. Glycogen Metab., Proc. Meet. Fed. Eur. Biochem. Soc. 4th* 1967 (Pub. 1968), 1-18 (Eng). Edited by Whelan, W. J. Acad. Press: London, Engl. Transferase, the enzyme that catalyzes the synthesis of the α-1,4-linkages of glycogen, is subject to control by several mechanisms, hormonal as well as nonhormonal. These controls occur rapidly in a no. of tissues which are sensitive to the actions of the hormones. An important biochem. mechanism consists of interconverting 2 forms of the enzyme with 2nd-stage interconverting enzymes. These catalyze the phosphorylation and dephosphorylation of the 2 forms of transferase. The site of the nonhormonal control by glycogen is identified as the phosphatase, while the site of the hormonal control by insulin and epinephrine is the kinase. Insulin acts at the kinase site to bring about a greater dependence on cyclic adenylate and thus inactivate the kinase, with no decrease in cyclic adenylate tissue concns. Epinephrine acts to increase tissue levels of cyclic adenylate and thus promote kinase action. RCTT

28501z **Branching enzymes.** Manners, David J. (Heriot-Watt Univ., Edinburgh, Scot.). *Contr. Glycogen Metab. Proc. Meet. Fed. Eur. Biochem. Soc. 4th* 1967 (Pub. 1968), 83-100 (Eng). Edited by Whelan, W. J. Acad. Press: London, Engl. The properties of branching enzymes from mammalian, yeast, bacterial, and plant tissues are compared. All of these enzymes can introduce addnl. α-(1 → 6)-glucosidic interchain linkages into amylopectin with the formation of a glycogen-type polysaccharide. The combined action of glycogen synthetase and a branching enzyme leads to the formation of a multiply branched polysaccharide. With mammalian liver and muscle branching enzymes, chains of about 7 glucose residues are detached from the outer chains of glycogen and are transferred to a (1 → 6)-position. The level of branching enzyme activity in fetal tissues and in human cases of amylopectinosis (glycogen-storage disease Type IV) is discussed. RCTT

28502a **Rabbit muscle amylo-1,6-glucosidase: properties and evidence of heterogeneity.** Taylor, Pamela M.; Whelan, W. J. (Sch. Med., Roy. Free. Hosp., London, Engl.). *Contr. Glycogen Metab., Proc. Meet. Fed. Eur. Biochem. Soc. 4th* 1967 (Pub. 1968), 101-14 (Eng). Edited by Whelan, W. J. Acad. Press: London, Engl. Muscle and liver amylo-1,6-glucosidases act on 6-α-glucosyl Schardinger dextrins to release glucose, and these substrates are specific for the assay of the enzyme in crude exts. These are the substrates of choice for investigation of glycogen-storage disease Type III (absence of glycogen debranching activity). Rabbit muscle amylo-1,6-glucosidase was examd. for its pH-activity profile with the α-glucosyl Schardinger dextrin substrate. The result differed from other reported observations based on hydrolysis of glycogen phosphorylase limit dextrin (φ-dextrin) and incorporation of glucose-^{14}C into glycogen. The differences between the pH profiles reported here and elsewhere are shown to be partly attributable to the fact that phosphate buffer inhibits the glucosidase activity at neutral pH values while glycylglycine inhibits on the acid side. Both buffers were employed in this and earlier studies. Passage of amylo-1,6-glucosidase through a DEAE-cellulose column caused a sepn. of the enzyme activity into fractions whose relative behaviors towards α-glucosyl Schardinger dextrins and to φ-dextrin were different. It is concluded that amylo-1,6-glucosidase is not a single enzymic species, but can be thought of as 2 enzymes, 1 with an acid and 1 with a neutral pH optimum. RCTT

28503b Enzymes in an insoluble phase. Variation of enzymic activity as a function of substrate composition. Regulators of enzymic membranes. Selegny, Eric; Broun, Georges; Thomas, Daniel (Lab. Chim. Macromol., Fac. Sci. Rouen, Mont-Saint-Aignan, Fr.) *C. R. Acad. Sci., Ser. D* 1969, 269(14), 1330-3 (Fr); cf. *CA* 71: 9881m. A math. equation was derived to describe the activity of an enzyme in an insol. phase. By developing a series using the formula of MacLaurin, the factors which condition the regulating stages of enzymic activity were detd. Exptl. results obtained with a synthetic membrane (50 μ thick) using glucose oxidase agreed with the theoretical predictions.
Dorothy J. Buchanan-Davidson

28504c Study of the thermal denaturation of ribonuclease by differential scanning calorimetry. Delben, F.; Crescenzi, V.; Quadrifoglio, F. (Univ. Trieste, Trieste, Italy). *Int. J. Protein Res.* 1969, 1(3), 145-9 (Eng). A study of thermal denaturation of RNase by differential scanning calorimetry in H_2O, in 1.5, 2.5, and $4M$ urea and in 1 and $2M$ aq. hexamethylenetetramine (II) indicated that urea lowers the m.p. of RNase but only slightly changes the enthalpy variation accompanying denaturation; I affected neither.
David B. Sabine

28505d Modified substrates and modified products as inducers of carbohydrates. Reese, Elwyn T.; Lola, J. E.; Parrish, Frederick W. (Pioneering Res. Lab., U.S. Army Natick Lab., Natick, Mass.). *J. Bacteriol.* 1969, 100(3), 1151-4 (Eng). Cellobiose and isomaltose are both inducers and repressors of cellulase and dextranase, resp. The repression can be avoided by supplying the disaccharide slowly. This was done by use of palmitate and acetate esters which are hydrolyzed by esterases of the growing organism to yield the inducer. Sucrase yields, also, are greatly increased (to 80-fold) by substituting sucrose monopalmitate for sucrose in culture media. RCJM

28506e Protein-protein interaction and the regulation of enzyme activity. Freiden, Carl (Sch. of Med., Washington Univ., St. Louis, Mo.). *Regul. Enzyme Activ. Allosteric Interactions,* *Proc. Meet. Fed. Eur. Biochem. Soc., 4th 1967* (Pub. 1968), 59-71 (Eng). Edited by Kvamme, E. Acad. Press: London, Engl. The ability of enzymes to undergo assocn.-dissocn. reactions appears to be of considerable importance with respect to metabolic control. What was proposed in this paper is that different mol. characteristics with respect to the rate of the assocn.-dissocn. reaction between like subunits of an enzyme may reflect different control functions in vivo. Rapid assocn.-dissocn. reactions may confer on an enzyme system an abnormal sensitivity to ligand concn. resulting in a sensitive point of metabolic regulation. A slow assocn. or dissocn. reaction, which frequently appears to interconvert active and inactive enzyme forms, may serve to act as a buffer to metabolic stress without altering the sensitivity to ligand concn. characteristic of the enzyme. RCTT

28507f Production of urokinase. Sloane, Nathan H. (Century Laboratories, Inc.) U.S. 3,477,913 (Cl. 195-66; C 07g), 11 Nov 1969, Appl. 31 Oct 1967; 5 pp. In the production of urokinase (I) a nucleo:otein-tannate ppt., prepd. by the addn. of tannic acid to a nucleoprotein, is added to human urine to adsorb I. The I-contg. ppt. is sepd. and I is dissolved by addn. of an alk. soln.; impurities are removed from the solubilized I. Thus, a nucleo-protein-tannate ppt. was prepd. by dissolving 50 mg human serum albumin and 50 mg purified yeast nucleic acid in 10 ml H_2O, adding excess tannic acid soln. (5 g tannic acid, 120 ml H_2O, 120 ml EtOH, and 2.5 ml HOAc), and mixing until pptn. was complete. The ppt., removed by centrifugation, was washed with distd. H_2O until free of acid, then suspended in H_2O. Three ml of the ppt. suspension was added to 125 ml of human urine and the mixt. was stirred for 1 hr at room temp. for complete adsorption of I. The I-contg. ppt. was removed by centrifugation, suspended in 7 ml of cold 0.05M Tris buffer (pH 7.4), and cold 0.2N NaOH was added dropwise to dissolve I at pH 9-10. During this procedure, the mixt. was cooled in an ice bath. I was dialyzed against 0.05M Tris buffer (pH 7.4) to remove excess alkali, tannic acid, and other impurities. The I conc. contained a total of 1250 CTA units (12,500 CTA units/l. urine). S. P. Marino

28288k **Computer literature bibliography. Volume II. 1964–1967.** Youden, W. W. (Inst. of Appl. Technol., Nat. Bur. of Stand., Washington, D.C.). *Nat. Bur. Stand. (U.S.), Spec. Publ.* **1968**, NBS Spec. Publ. 309, 381 pp. (Eng). Avail. GPO. $5. This bibliography is a continuation of *Computer Literature Bibliography, 1946 to 1963*, and is intended as a further service to the computer community. It contains approx. 5200 references to computer literature published during the years 1964 through 1967. The Bibliography Section includes the full title and the names of all of the authors of each item published in 17 journals, 20 books composed of chapters by individual authors, and 43 conference proceedings. In addn., refs. to all items that were reviewed in the IEEE Transactions on Electronic Computers have been included. The Title Word Index Section provides the means for locating an item if any part of its title is known. Likewise, it can be used to identify all items whose titles include a particular word or phrase. The Author Index Section lists all authors of each item, but does not indicate whether an individual is its sole author. RCTT

28289m **Automatic documentation. Principles and application. Automation of a large number of documents. Projects and accomplishments of the [Swiss] Federal Institute of Technology library.** Sydler, J. P. (EPF, Zurich, Switz.). *Neue Tech. A* **1969**, 11(5), 269–75 (Fr). The projects and accomplishments of the Swiss Federal Institute of Technology Library (planned automatic documentation, automatization of loans and cataloguing, solution for input difficulties and importance of the output, and systematic thesaurus) are discussed. Shanta Baloo

28290e **Application of the SAGESSE system in a large electronic project.** Gamp, J.; Tamman, L. (CSF, Paris, Fr.). *Neue Tech. A* **1969**, 11(5), 275–80 (Fr). Collaboration among the Centre of Documentation, its computation center, and the IBM Company led to the SAGESSE system; it is applied in a large electronic project. The first tests were successful in 1964, and since 1966 the propagation of automatic documentation covers all labs and works of the Sans Fil Company (telegraph). Shanta Baloo

28291f **Atomic time keeping at the National Research Council.** Mungall, A. G.; Daams, H.; Bailey, R. (Div. Appl. Phys., Nat. Res. Coun. Canada, Ottawa, Can.). *Metrologia* **1969**, 5(3), 73–6 (Eng). Three different exptl. at. time scales maintained for the past 4 years at the National Research Council of Canada (NRC) are described. One is based on an HP 5061A Cs standard (*1*), another on a free-running crystal oscillator calibrated daily with respect to the NRC 2.1 m primary Cs standard (*2*), and a 3rd one on the HP 5061A calibrated weekly by the primary standard (*3*). Performances of the above time scales are intercompared. No. *3* is found to be the most accurate on the basis of VLF and Loran C intercomparison with USN observatory and NBS (Natl. Bureau of Standards) scales. Agreement between *3* and those of USNO and NBS is of the order of 1 μsec or 1 × 10[13] in frequency over a 4-month period. K. K. Desai

28292g **Neutron activation analysis of potsherds from Knossos and Mycenae.** Harbottle, German (Brookhaven Nat. Lab., Upton, N.Y.). *U.S. At. Energy Comm.* **1968**, BNL-13740, 24 pp. (Eng). Avail. Dep.; CFSTI. From *Nucl. Sci. Abstr.* **1969**, 23(17), 33159. Twenty potsherds of each of 2 groups, representing pottery finds at Mycenae and Knossos, are analyzed by neutron activation for 10 elements. The results indicate that this procedure can be a useful complement to the emission spectrographic method of anal., both as a check on the latter and as a means of extending the range of elements determinable to considerably lower concns. One sample in each of the 2 groups definitely seems not to belong to its supposed compositional category: with these exceptions, the groups established by emission spectrographic anal. appear to retain their homogeneity when elements present at a lower concn. level are detd. TCNG

28293h **Biology and the Physical Sciences.** Devons, Samuel; Editor (Columbia Univ. Press: New York). **1969**. 379 pp. $12.50. Textbook.

28294j **Electron Microscopic Methods (Elektronenmikroskopische Methodik).** Schimmel, G. (Springer-Verlag: Berlin). **1969**. 243 pp. 78 DM.

28295k **Ullmann's Encyclopedia of Technical Chemistry, Vol. 19: Cement to Intermediate Products. 3rd ed.** (Ullmanns Encyclopaedie der Technischen Chemie, Bd. 19: Zement-Zwischenprodukte). Foerst, W.; Editor (Urban and Schwarzenberg: Munich, Ger.). **1969**. 418 pp. Reviewed in *Chimia* **1969**, 23(10), 375 (Ger).

28296m **The Use of Chemical Literature (Information Sources for Research and Development). 2nd ed.** Bottle, R. T; Editor (Butterworth: London). **1969**. 294 pp. 65s.

See also: Exptl. use of punch cards in studying the sanitary conditions of discharging waste waters into reservoirs (Sect. 60) **24341z.** Investigation of combined patterns from diverse anal. data using computerized learning machines (Sect. 79) **27958s.** Computer program for calcg. corrections for electron microprobe anal. (Sect. 79) **27956q.**

2—GENERAL BIOCHEMISTRY

G. FRED SOMERS AND EDITOR EMERITUS L. E. GILSON

28297n Atomic energy in agricultural research. Haq, M. Shamsul *Agr. Pak.* 1968, 19(3), 461–74 (Eng). This is a review. The application in Pakistan of at. energy to research on plant breeding and genetics, plant physiol., soil science, liming of acid soils, rice fertilization, and entomol. is discussed.
Kurt H. Zingraf

28298p Escherichia coli ribosomes: recent developments. Schlessinger, David; Apirion, David (Sch. of Med., Washington Univ., St. Louis, Mo.). *Annu. Rev. Microbiol.* 1969, 23, 387–426 (Eng). A review. Both 30S and 50S ribosomes seem to be nearly homogeneous and have complex structures formed from 1 or 2 RNA mols. and 20 or 30 protein mols. Precursors of both ribosomes were found by pulse labeling and their structure detd. Protein synthesis is carried out by a 70S monosome formed during initiation by the binding of first a 30S and then a 50S ribosome to the messenger RNA. Upon termination of the synthesis of a protein the monosome dissocs. to its 50S and 30S subunits. At least 3 factors required for this cycle were isolated. Six different loci of antibiotic resistance were mapped and studied. A few other types of mutants which affect ribosomes are known. 238 refs.
D. B. Wilson

28299q Synthetic and division rates of Euglena: a comparison with metazoan cells. Wilson, Barry William; Levedahl, Blaine H. (Univ. of California, Davis, Calif.). *Biol. Euglena* 1968, 1, 315–32 (Eng). Edited by Buetow, Dennis E. Acad. Press: New York, N.Y. A review comparing cell compn., synthetic rates, and division rates in *Euglena* with those of metazoan cells. 13 refs.
L. K. Allen

28300h Bionics. Greguss, Pal *Fiz. Szemle* 1969, 19(8), 225–34 (Hung). Bioelectronic phenomena with special emphasis on electric fishes are reviewed. 9 refs.
A. C. Pronay

28301j Calculation of polypeptide conformation. Scheraga, Harold A. (Cornell Univ., Ithaca, N.Y.). *Harvey Lect.* 1967–1968 (Pub. 1969), Ser. 63, 99–138 (Eng). A review of the development of a method of calcg. polypeptide conformations through use of the lowest free energy values. 76 refs.
Edward J. Hennessy

28302k Structure and function of hemoglobin. Perutz, Max F. (Med. Res. Counc. Lab. Mol. Biol., Cambridge, Engl.). *Harvey Lect.* 1967–1968 (Pub. 1969), Ser. 63, 213–61 (Eng). The structure and function of Hb are extensively discussed. The structural relation of deoxyhemoglobin and oxyhemoglobin is described. Electron d. maps of horse oxyhemoglobin and its amino acid sequence are presented, as well as stereoscopic drawings of the atomic model of the mol. Polar and nonpolar interactions, dissocn., and effects of the heme group are discussed. 57 refs.
F. S. Carter

28303m Genome of bacteriophage T4. Edgar, R. S. (California Inst. of Technol. Pasadena, Calif.). *Harvey Lect.* 1967–1968 (Pub. 1969), Ser. 63, 263–81 (Eng). A discussion is given of the genome of bacteriophage T4 as related to gene functions. A gene map of T4 is given. The functions of the genes are listed and discussed with respect to translation and transcription of viral DNA, breakdown of host DNA, synthesis of 5-hydroxymethylcytosine, replication and assembly of the phage, and effects on host cell membrane. 37 refs.
F. S. Carter

28304n Bacterial cell wall analysis. I. Cell wall preparation. Drucker, D. B. (Dent. Sch., Univ. Wales, Cardiff, Wales). *Lab. Pract.* 1969, 18(11), 1181–4 (Eng). Research undertaken in the study of bacterial cell walls in the last 20 years and the prepn. of material for anal. are reviewed. Prepn. includes disruption (osmolysis, thermal rupture, autolysis, solid shear, and liq. shear), cell-wall purification, and hydrolysis. 5 references.
V. N. Nekrassoff

28305p Molecular bonding in biology: theory and experience. Pullman, Bernard (Inst. Biol. Phys.-Chim., Paris, Fr.). *Mol. Ass. Biol., Proc. Int. Symp.* 1967 (Pub. 1968), 1–19 (Fr). Edited by Pullman, Bernard. Acad. Press: New York, N.Y. A discussion of mol. assocns. in biol. via H bonding in theory and experience is presented. Intermol. forces involved and interaction energies are reported for various biomols. 37 references.
William Braker

Glossary of Selected Terms and Abbreviations

ACS
: American Chemical Society.

ASTIA
: *See* DDC.

ASTM
: American Society for Testing and Materials, an organization which (among other activities) promulgates standards in many fields of science and technology.

AUTHOR ABSTRACT
: An abstract written by the author of the original document.

AUTOMATIC INDEXING
: Indexing by computer, often based on full text of original document or of abstract.

BA
: *Biological Abstracts*, a major abstracting service in biology.

BIOSIS
: Biosciences Information Service of *Biological Abstracts*.

CA
: *Chemical Abstracts,* a major abstracting service in chemistry.

CAC
: *Current Abstracts of Chemistry*, a publication of the Institute for Scientific Information.

CAS
> Chemical Abstracts Service, the "parent" organization for *CA* and other related services. It is a Division of the American Chemical Society.

CATHODE RAY TUBE
> A television-like device, which can display information (such as abstracts) when linked to a computer-based system.

CODEN
> A five character identification for periodical titles as published by the American Society for Testing and Materials (ASTM).

COMPUTER-BASED
> Describes a system or process (or publication) that functions (or is issued) through the use of a computer.

COMPUTER PROGRAM
> A series of instructions that "tell" a computer how to process data.

COMPUTER-READABLE
> *See* Machine-readable.

CRITICAL ABSTRACT
> An abstract which contains an evaluation or critique of the original document.

CRT
> *See* Cathode ray tube.

CUMULATIVE INDEX
> An index which is usually based on the merging of several annual, semi-annual, or monthly indexes.

CURRENT ALERTING ABSTRACTING SERVICE
> As used in this book, an abstracting service that places great emphasis on speed of issue.

CURRENT AWARENESS
> Used by many people as synonymous with SDI (*see also* SDI).

DATA BASE
> An extensive, organized, indexed collection of abstracts (or citations to abstracts) that can be processed and searched by computer.

DDC
> Defense Documentation Center, an agency that is a major information center for the United States Department of Defense.

EDP
> Electronic data processing (which usually means processing of data using a computer).

EI

Engineering Index, a major abstracting service in engineering.

EXPRESS SERVICE

See Current alerting abstracting service.

FEEDBACK

User or client response or opinion.

"FLEXOWRITER"

A device which "captures" typing on punched paper tape for ease of revision or for other uses. A product of the Friden Division of the Singer Company.

"FRIDEN 2340"

A device similar to the "Flexowriter" made by the same manufacturer, but with some features different from the "Flexowriter."

GRAPHICS

Nonverbal material, e.g., diagrams, drawings, photographs, graphs.

HIGHLIGHT ABSTRACTS

As used in this book, abstracts used by editors of journals to inform readers more fully about the contents of articles in the journal issues.

IBM

International Business Machines Corporation.

IBM "360/30"

One of a "family" of IBM computers.

INDICATIVE ABSTRACT

A brief abstract that tells what a document is about.

INFORMATION CENTER

A department or function that collects, processes, evaluates, summarizes, and disseminates information in direct interaction with scientists and engineers. The staff often includes information scientists, librarians, and computer specialists.

INFORMATION SCIENTIST

A trained scientist whose primary current experience and assignment relates to the literature and to information handling.

INFORMATIVE ABSTRACT

A full-length abstract that contains the essential data and conclusions of a document.

IN-HOUSE

Refers to an activity conducted within an organization, the product of which activity is usually not made available to other organizations or to the general public.

INSPEC

Information Service in Physics, Electrotechnology and Computers and Control. The Service publishes current awareness and abstract journals in these fields. Published by the Institution of Electrical Engineers in association with a number of other technical societies.

INTEREST PROFILE

The listing of the specific areas or fields of interest to a specific individual as intended for use in SDI (*see also* SDI).

KEYBOARD (verb)

To type or otherwise similarly record data or information usually for direct subsequent computer input and processing.

KEYWORDS

Words or terms used to index a document; sometimes synonymous with "subject index entries."

LITERATURE SEARCHER

See Information scientist.

MACHINE-READABLE

Data (letters and/or numbers) in a physical form that can be read (understood) by computers and related equipment.

MAGNETIC TAPE

A plastic tape coated with a magnetic material on which information can be recorded or read (understood) by computers or computer-related devices.

"MAGNETIC TAPE SELECTRIC® TYPEWRITER"

An IBM device which "captures" typing on magnetic tape for ease of revision and other uses.

MECHANIZED

See Computer-based.

METAL OXIDE

Magnetic material used to coat magnetic tape.

MICRO-ABSTRACT

See Mini-abstract.

MICROFICHE

A type of microform consisting of a film sheet usually with approximate dimensions of 4″ × 6″; this size can accommodate 60 images reduced 20 to 1 and arranged in 5 horizontal rows or 98 images reduced 24 to 1 in 7 rows. In addition, there are other configurations in use.

MINI-ABSTRACT
An especially brief indicative abstract.

"MTST"
An IBM product; *see* "Magnetic Tape Selectric® Typewriter."

NFSAIS
National Federation of Science Abstracting and Indexing Services.

ON-LINE DEVICE
Device for data input/output directly connected to a computer.

ORIENT (verb)
As used in this text, to write abstracts to meet needs of a specific audience.

PRIMARY JOURNAL
A journal which publishes papers describing the results of original research.

PROFILE
See Interest profile.

PROGRAM
See Computer program.

PROGRAMMER
A person skilled in the writing of computer programs.

PROPRIETARY
When applied to information, that which is private or confidential to a specific organization.

REFEREE
A scientist who assists a journal editor in evaluating the acceptability of papers submitted for publication.

REMOTE TERMINAL
An on-line device (such as a typewriter) directly connected to a computer but physically separated from it. This separation could be a few yards or it could be miles (*see also* On-line).

RETROSPECTIVE
Noncurrent, e.g., "searching of noncurrent literature."

SATCOM
Acronym for Committee on Scientific and Technical Communication of the National Academy of Sciences—National Academy of Engineering.

SDI
> Selective Dissemination of Information to specific individuals as based on their specific interests.

SECONDARY SERVICE
> Sometimes used instead of "Abstracting Service" with which it is usually synonymous.

SLANT
> *See* Orient.

TAPE
> *See* Magnetic tape.

TERMINAL
> *See* Remote terminal.

THESAURUS
> A list of subject headings including cross-references (such as "see" and "see also") intended for use as a guide to indexers and index users.

TRADE JARGON
> The "slang" of a specific profession or field.

VIDEO DISPLAY
> Display of information on a cathode ray tube or similar device. (*See also* Cathode ray tube.)

References

1. "ACS Report Rates Information System Efficiency," *Chemical and Engineering News*, **47**, July 28, 1969, pp. 45–46.
2. "CA Prints 4-Millionth Abstract," *Chemical and Engineering News*, **46**, August 12, 1968, p. 44.
3. *News from Science Abstracting and Indexing Services*, **11**, No. 6, December 1969, p. 5.
4. R. E. Maizell, "Information Gathering Patterns and Creativity," *American Documentation*, **11**, January 1960, pp. 9–17.
5. A. Uris, "How to be a Great Dictator," *Chemical Engineering*, **76**, No. 14, June 30, 1969, pp. 111–113.
6. J. R. Fair, "Dictation and the Engineer," *Chemical Engineering*, **76**, No. 14, June 30, 1969, pp. 114–117.
7. P. D. Leedy, *A Key to Better Reading*, McGraw-Hill, New York, 1968.
8. *Science Information Notes*, **10**, August-September 1968, p. 10.
9. R. Calvert, editor, *The Encyclopedia of Patent Practice and Invention Management*, Reinhold, New York, 1964.
10. *Kirk-Othmer, Encyclopedia of Chemical Technology*, 2nd ed., Vol. **14**, Wiley, New York, 1967, pp. 552–635.
11. *Handbook for Authors of Papers in the Journals of the American Chemical Society*, American Chemical Society, Washington, 1967, pp. 20–22.
12. F. W. Lancaster, *Information Retrieval Systems*, Wiley, New York, 1968, pp. 162–165.
13. *Standard Industrial Classification Manual*, Executive Office of the President, Bureau of the Budget, U.S. Government Printing Office, Washington, 1967.
14. F. W. Lancaster, *Information Retrieval Systems*, Wiley, New York, 1968, p. 37.

15. F. W. Lancaster, *Information Retrieval Systems,* Wiley, New York, 1968.
16. Harold Hart, "Articles for Chemical Reviews—Suggestions to Authors," *Chemical Reviews,* **67,** No. 4, July 25, 1967, pp. 361–366.
17. J. H. Kuney and S. Anderson, "Industrial and Engineering Chemistry Research Results Service," *Journal of Chemical Documentation,* **5,** August 1965, pp. 128–131.
18. J. J. Magnino, Jr., "IBM Technical Information Retrieval Center—Normal Text Techniques," ITIRC-002, Technical Report, April 1, 1965; "IBM's Unique but Operational International Industrial Textual Documentation System—ITIRC," ITIRC-023, Technical Report, September 1967; "Normal Text Information Retrieval—Management and Line Use within IBM," ITIRC-017, Technical Report, May 1967. IBM, Armonk, New York.
19. G. O. Walter, "Typesetting," *Scientific American,* **220,** May 1969, pp. 60–69.
20. F. A. Tate, "Progress Toward a Computer-Based Chemical Information System," *Chemical and Engineering News,* **45,** January 23, 1967, pp. 78–90.
21. R. J. Rowlett, Jr., F. A. Tate, and J. L. Wood, "Relationships between Primary Publications and Secondary Information Services," *Journal of Chemical Documentation,* **10,** February 1970, pp. 32–37.
22. J. H. Kuney, "New Developments in Primary Journal Publication," *Journal of Chemical Documentation,* **10,** February 1970, pp. 42–46.

Bibliography

D. B. Baker, "Communication or Chaos," *Science,* **169,** pp. 739–742 (August 21, 1970).

J. B. Bennett, *Editing for Engineers,* Wiley, New York, 1970.

K. D. Carroll, Ed., *Survey of Scientific-Technical Tape Services,* American Institute of Physics, New York, and American Society for Information Science, Washington, 1970.

L. Cohan, Ed., *Directory of Computerized Information in Science and Technology,* Science Associates/International, New York, 1970.

Committee on Scientific and Technical Communication, *Scientific and Technical Communication—A Pressing National Problem and Recommendations for Its Solution,* National Academy of Sciences, Washington, 1969.

R. H. Dodds, *Writing for Technical and Business Magazines,* Wiley, New York, 1969.

E. Ehrlich and D. Murphy, *The Art of Technical Writing,* Thomas Y. Crowell, New York, 1964.

G. Jahoda, *Information Storage and Retrieval Systems for Individual Researchers,* Wiley, New York, 1970.

J. H. Mitchell, *Writing for Professional and Technical Journals,* Wiley, New York, 1968.

National Science Foundation, *Current Research and Development in Scientific Documentation,* U. S. Superintendent of Documents, Washington, 1969.

G. Salton, *Automatic Information Organization and Retrieval,* McGraw-Hill, New York, 1968.

G. Salton, "Automatic Text Analysis," *Science,* **168,** pp. 335–343 (April 17, 1970).

B. H. Weil, "Standards for Writing Abstracts," *Journal of the American Society for Information Science,* **21,** pp. 351–357 (No. 5, September-October, 1970).

F. P. Woodford, Ed., *Scientific Writing for Graduate Students—A Manual on the Teaching of Scientific Writing,* The Rockefeller University Press, New York, 1968.

Index